IØ221574

Alexander Winchell

The Grand Traverse Region

A report on the geological and industrial resources of the counties of

Antrim, Grand Traverse, Benzie and Leelenaw in the lower peninsula of

Michigan

Alexander Winchell

The Grand Traverse Region
A report on the geological and industrial resources of the counties of Antrim, Grand Traverse, Benzie and Leelenaw in the lower peninsula of Michigan

ISBN/EAN: 9783337309053

Printed in Europe, USA, Canada, Australia, Japan

Cover: Foto ©berggeist007 / pixelio.de

More available books at **www.hansebooks.com**

THE GRAND TRAVERSE REGION.

A REPORT

ON THE

GEOLOGICAL AND INDUSTRIAL RESOURCES

OF THE COUNTIES OF

ANTRIM, GRAND TRAVERSE, BENZIE AND LEELANAW

IN THE

LOWER PENINSULA OF MICHIGAN.

By ALEXANDER WINCHELL, A. M.,

Prof. of Geology, Zoology and Botany in the University of Michigan, and late State Geologist ; Member of the Geological Society of France ; Corresponding Member of the Geological Society of Glasgow ; Member of the American Philosophical Society ; Corresponding Member Boston Society of Natural History, &c.

ANN ARBOR:

DR. CHASE'S STEAM PRINTING HOUSE.

1866.

PREFACE.

The following report has been drawn up for the purpose of directing attention to the most remarkable and desirable section of country in the Northwest. Emigrants and capitalists will equally find in it statements of facts which will both surprise and interest them. I have no fear of being charged with overdrawing the picture. I have only given facts, figures and vouchers. They speak for themselves. The details of the geology of the region have never before been worked out, and will prove of interest to a large class of readers.

This region, like all of Northern Michigan, has heretofore been generally misrepresented. I gladly except from this charge the account drawn up by Hon. D. C. Leach for *Clark's Gazetteer of Michigan;* an interesting and extended statement which appeared in the *Grand Traverse Herald* for March 4th, 1864, and the address of D. B. Duffield, Esq., on the "Undeveloped Regions and Resources of the State of Michigan," as well as the various pamphlets and essays of Edgar Conkling, Esq., of Cincinnati, and George S. Frost, Esq., of Detroit.

In the prosecution of my work I have been greatly aided by the courtesies and liberality of the citizens of the region—especially of some whom I would be glad to name if justice only had to be consulted in the matter. I was accompanied during more than half of my explorations by A. de Belloy, Esq., of Sutton's Bay, who will be glad to reply to any inquiries respecting the region. I cannot forbear to mention the great aid I have received from the exact and reliable maps of S. Farmer & Co., of Detroit. Farmer's Sectional Map has been my pocket companion in all my travels in various parts of the State for the past eight years, and I have learned to rely upon it implicitly even to the obscurest highways and the meanderings of the smallest creeks. Many of my meteorological data have been taken from the published observations of the Smithsonian Institution. The original and published maps of the Lake Survey have been freely submitted to my inspection through the courtesy of Colonel W. F. Raynolds, Superintendent, and I am indebted to A. S. Packard, Jr., M. D. of Boston, for the identification of the canker worm moth of the region.

A. WINCHELL.

UNIVERSITY OF MICHIGAN,
    Ann Arbor, December, 1865.

THE GRAND TRAVERSE REGION.

I. THE NAME.

The early French *voyageurs* in coasting from Mackinac southward found two considerable indentations of the coast line of Lake Michigan on the east side, which they were accustomed to cross from headland to headland. The smaller of these they designated "La Petite Traverse" and the greater, "La Grande Traverse." These names were transferred to the two bays known as the Little Traverse and Grand Traverse Bays.

II. GEOGRAPHICAL POSITION.

Grand Traverse Bay is a bay of lake Michigan, indenting the northwestern shore of the southern Peninsula of the State of Michigan. Its general direction is from north to south. Its mouth is in latitude 45° 15' north, and its head in latitude 44° 45' north. Its length in a straight line is therefore 34.75 statute miles. The undefined region bordering on this bay is generally known as the Grand Traverse Region. The county of Antrim lies upon the east side of the bay, the county of Leelanaw on the west, and the county of Grand Traverse on and about the head of the bay. The counties of Benzie on the west and Kalkasca on the east of Grand Traverse, may be regarded as lying within the same topographical and hydrographical area; and in their geological and physical features belong to the same district.

Leelanaw county occupies the triangle lying between Grand Traverse Bay and lake Michigan. Grand Traverse county embraces the tongue of land which bisects the southern half of the bay, and extends northward to embrace about nine miles along the eastern shore of the bay. Benzie county lies upon lake Michigan. Kalkasca county is reached by navigable water only in the northwestern corner, through Elk and Round lakes. The southern limit of the region thus indicated lies in latitude 44° 30' and the northern limit in latitude 45° 15' north.

III. HYDROGRAPHY.

Grand Traverse Bay is a sheet of navigable water about thirty-three miles in length with an average breadth of about eleven miles. The southern portion of the bay is divided into the east and west arms by a belt of land from one to two miles wide and about seventeen miles in length, known as "the Peninsula." The east arm has an average width of about four and a half miles; the west arm is somewhat wider. The height of the bay and of lake Michigan above the level of the sea is 578 feet. The depth of water in the bay is generally from 20 to 70 fathoms. The east arm attains the greatest depth, being about a hundred fathoms at a point opposite Old Mission and thence as far as Petobego Lake. The maximum depth is 618 feet, and is found opposite Birch Lake and on a line between Old Mission and the north end of Elk lake.

The entire bay constitutes a harbor secure from all except northerly winds; while the two arms of the bay are not seriously disturbed by storms from any direction. The shores of the bay however, present a number of harbors in which vessels may at all times lie with the utmost security. Entering the bay at its mouth and proceeding along the western shore, the first important harbor reached is Northport which opens towards the south—being separated from the bay by a tongue of land called "Carrying Point." This harbor is about two miles wide and nearly three miles deep and is a frequent resort of vessels overtaken by storms upon the lake. The water is sufficient for the largest vessels which navigate the lakes.

Proceeding southward, twelve miles from the mouth of the
bay we reach New Mission Harbor, also opening southward
and separated from the bay by Shobwasson Point. This harbor is a mile and a half wide and a mile deep, with an abundance of water for safe navigation.

Four miles further south is Sutton's Bay, opening towards
the northeast, and separated from the West Arm by Stony
Point. This harbor is three miles long and a mile and a half
wide with plenty of water.

Lee's Point, eleven miles from the head of the West Arm
forms another shallow harbor.

Bower's harbor, on the west side of the Peninsula, opens to
the southwest, being isolated from the West Arm by Tucker's
Point. Off this point, and connected with it by a reef, is Harbor Island—practically extending Bower's harbor to the length
of over three miles, while its width is about one and a half
miles.

On the east side of the Peninsula, near the point, is Old
Mission harbor, having a capacity of about one square mile.
Further than this the configuration of the shore of the East
Bay affords no harbor worthy of note.

Passing southward from the mouth of Grand Traverse Bay
along the shore of lake Michigan, we find a broad indentation
at the mouth of Carp River, opening towards the northwest
and partially protected from west and southwest winds by
Mount Carp.

Between Mount Carp and North Unity is a broad bay about
five miles deep, affording protection from all winds except those
proceeding from the north and northwest.

Between North Unity and Sleeping Bear Point is another
broad bay about four miles deep, forming the harbor of Glen
Arbor, affording shelter from all except north and northwest
winds. The mouth of this harbor opening towards the Manitou Islands about nine miles distant, it receives considerable
protection from heavy " seas " approaching from that quarter.

No other natural harbor of importance exists along this
shore ; though improvements, some of which are now in pro-

gress—as at the mouth of the Bees Scies River—will create harbors equal in excellence to any in the region.

The Grand Traverse region is remarkably provided with navigable inland lakes. Some of these connect with each other or with the bay or lake Michigan in such a manner as to constitute extended channels of inland communication by water. Connecting with the East Arm of the bay through Elk river is Elk lake, a body of water about ten miles long and averaging a mile and a half in width. Passing from this we enter Round lake, about one-fifth as large, from which we proceed northward to Torch lake, the largest in the region. This sheet of water is eighteen miles long, and averages about two miles in width. It lies nearly parallel with the east shore of the bay, the upper extremity approaching within half a mile of the latter. From the east side of Torch lake we pass into Clam lake, a narrow strip of water stretching eastward into Grass lake. From the latter we proceed through a series of small lakes extending northward about twelve miles, and called collectively Intermediate lake. The upper extremity of this lake is but two miles from the south arm of Pine lake, lying mostly in Emmet county and discharging through Pine river into lake Michigan.

The remarkable series of lakes just described is navigable for tugs and small vessels from the East Arm of the bay to the head of Grass lake, making a total length of navigable inland water amounting to eighty miles. Pine lake affords about forty-two miles of inland navigation.

Carp lake in Leelanaw county affords a stretch of inland navigation for tugs, amounting to thirty miles. The outlet of this lake is through Carp river. It extends nearly north and south with a mean width of nearly a mile—taking no account of "the narrows," where, for the distance of two miles the mean width is less than a quarter of a mile.

Glen lake in the same county, lies within one mile of lake Michigan with which it connects through Crystal creek. This body of water covers about one sixth of a township. It is over 200 feet deep—a depth of 15 feet being reached at the distance of ten rods from the shore. It is therefore navigable for vessels of large size, though none above twenty tons burden could approach through Crystal creek; and even this would necessitate some improvements.

Platte lake in Benzie county is similarly circumstanced, though smaller, and lying somewhat further from lake Michigan.

Lake aux Bees Scies (or lake " Betsie")—sometimes called Crystal lake—is the second in size of the lakes of this region. Its western extremity approaches within half a mile of lake Michigan, but its outlet is through the Bees Scies river and Frankfort harbor. The latter is a beautiful little lake reaching to within a few rods of lake Michigan with which it connects through the Bees Scies river. The improvements in progress here will render this harbor accessible from lake Michigan for any vessels which navigate the lake, and it will be absolutely secure from storms from any quarter of the compass. This harbor is two miles long and from a quarter to a half a mile in width.

Numerous smaller lakes of less importance dot the entire region, of which Petobego lake in the northeastern part, and Silver, Long, Bass and Green lakes in the western part of Grand Traverse county are beautiful sheets of limpid water with hard shores. Boardman's lake, within half a mile of Traverse City, is destined, in time, to afford a large accession to the sources of pleasure to the future population of that place. Lime and Bass lakes in the western part of Leelanaw county, Cedar lake in the southeastern part, and Leg lake in the northern part are among the smaller bodies of water with which the region is supplied.

These numerous lakes are filled with pure and palatable water; their shores are dry, and in connection with the surrounding scenery, they constitute, in addition to the facilities they afford for internal communication, the completion of the charms of a series of the most charming landscapes.

The streams of the region are naturally of inconsiderable magnitude. The Manistee river flows through the southeastern portion of Kalkasca county, and passes beyond the limits of the present notice. Boardman's river rises in the northern part of Kalkasca county, and, after flowing southwest about thirty miles, bends northward and flows about nine miles into the West Arm of Grand Traverse bay. Elk river, the outlet of Elk lake, is scarcely a quarter of a mile long. It discharges a large body of water, and has a sufficient fall to afford a first class water power. The river has been dammed by which the approach from the bay is cut off, while the depth of water in the chain of lakes lying towards the interior is proportionally improved for the purpose of navigation. Carp river, the outlet of Carp lake discharges a body of water nearly as large, and having a fall of five or six feet affords another admirable water power. Here also is a dam. This river is not over half a mile in length. Crystal creek, the outlet o Glen lake, is of smaller dimentions, pursuing a tortuous course of about three miles, and affording by its fall one or two good water powers. The Bees Scies river rises in a chain of lakes in the western part of Grand Traverse county, flows southwest about twelve miles, then north and west about eighteen miles to Frankfort harbor, through which it empties into lake Michigan. This stream affords a water power which is improved near Benzonia.

The Manistee, Boardman and Bees Scies rivers afford good mill sites in the unsettled regions through which their upper waters flow ; while numbers of smaller streams have been employed or may be, for driving mills to accommodate their immediate neighborhoods.

The streams of this region are supplied with pure clear water and flow with a lively current over pebbly bottoms to their places of discharge. There are very few instances o water colored by vegetable or peaty accumulations, or stagnated by flats, in the vicinity of the mouths of the streams.

Small brooks and rills are very numerous throughout nearly all parts of the region, so that there is scarcely a quarter section of land that is not supplied with living water, or that has not access to some of the numerous lakes with which the country is so abundantly supplied.

It will at once be noticed that this region is favored with an extent of navigable water which is quite remarkable. Not only is the whole extent of shore line of lake Michigan and the bay accessible for vessels of large draft, but to augment this shore line to a still greater extent, the bay is parted longitudinally for the distance of seventeen miles, and nearly the entire coast of the lake and bay is diversified by alternate " points " and indentations, which materially increase the means of access to the land. The whole extent of coast line bordering on lake Michigan is not less than seventy-five miles, of which fifty lie within Leelanaw county. Grand Traverse bay presents a coast line of 113 miles, of which 41 lie within Leelanaw county, 50 in Grand Traverse, and 23 in Antrim county.

The shore line of navigable water afforded by the various inland lakes is as follows : In Leelanaw county Carp lake affords about 36 miles and Glen lake about 14. In Benzie county Crystal lake affords about 20 miles and Frankfort harbor about 5 miles. In Antrim county and extending into Grand Traverse, Elk lake affords about 23 miles, Round lake 7 miles, Torch lake · 36 miles, Clam and Grass lakes 17 miles. Omitting mention of the considerable lakes in the western part of Grand Traverse county we thus have in this region 158 miles of shore line bordering on the navigable inland lakes, and 189 miles bordering on the bay and lake Michigan. This gives a total of 347 miles of shore line bordering on navigable waters in the region under consideration, and distributed as follows :

In Leelanaw county.............................141 miles
In Benzie county................................. 50 "
In Grand Traverse county........................ 63 "
In Antrim county................................ 93 "

    Total...................... ....347 "

IV. TOPOGRAPHY.

The mean elevation of the Grand Traverse region above lake Michigan may be estimated at 230 feet, or 808 feet above the level of the sea. The mean elevation of the lower Peninsula of Michigan is estimated by Higgins to be 160 feet above lake Michigan or 738 feet above the sea level.

The surface of the Grand Traverse region is thus seen to be somewhat elevated. Its configuration is undulating or broken. The drainage is almost perfect, so that swamps and stagnant waters are rarely encountered. The region on the west side of the bay is more uneven than that on the east. An elevated and somewhat broken tract extends from Lighthouse Point through Leelanaw and Benzie counties to beyond Frankfort. Back from this belt the country is equally elevated but less broken. Grand Traverse county is quite diversified with valleys, slopes and plateaux, but the surface rarely sinks so low or becomes so level as to interfere with complete drainage. The surface of Antrim county is undulating, sometimes hilly, and, though well watered, no marshes of importance occur.

Some parts of Leelanaw county present hills of somewhat formidable magnitude. Most of the northern part of the triangle is decidedly rough. The ridge of land separating Carp lake from Sutton's bay attains an elevation of nearly 400 feet above the bay. The slopes, however, are passable for loaded wagons. Carp lake is a beautiful sheet of pure water, resting in the bosom of the hills, which, with their rounded forest-covered forms, furnish it a setting of surpassing loveliness. Except for a short space on the east side, south of the narrows, the shores of the lake are occupied by dry and arable land. The region between Glen Arbor and Traverse City is substantially an undulating plateau lying at an elevation of about 300 feet above the lake. Glen lake is surrounded by hills, which attain an elevation of 250 to 400 feet. North Unity is a bold bluff of clay and sand, formed by the wasting of the lakeward side of a prominent hill by the action of the waves. Sleeping Bear Point is an enormous pile of gravel, sand and clay, which has been worn away on its exposed borders till the lakeward face presents a precipitous slope rising from the waters to an elevation of 500 feet, and forming with the horizon an angle of fifty degrees. Back from the face of the bluff is an undulating plateau of clay, pebbles and sand, covering an area of six or eight square miles, over which the only signs of vegetation are a few tufts of brown, coarse grass with scattered clumps of dwarfed and gnarly specimens of the balm of Gilead—a miniature desert, lying 380 feet above the lake. Across this waste of sand and clay the wind sweeps almost incessantly, —sometimes with relentless fury—driving pebbles and sand into the shelter of the neighboring forest, and causing the stunted poplars to shrink away in terror at its violence. The pelting sand has polished the exposed surfaces of the larger fragments of rocks to such an extent that they reflect the sunlight like a mirror. Their surfaces are sometimes worked into furrows, pits and grotesque inequalities in consequence of the unequal hardness of different portions of the stone. The "Bear" proper is an isolated mound rising a hundred feet above this desolate plateau and singularly covered with evergreens and other trees, presenting from the lake the dark appearance which suggested to the early navigators the idea of a bear in repose.

Empire bluff, six miles further south, presents a section of another hill which attains an altitude of nearly 400 feet, and the hills at Point Becs Scies reach an elevation but little less.

Seen from the lake, the natural cuts presented between Cathead Point and Carp river, at North Unity, Sleeping Bear and Point Becs Scies look like huge accumulations of blown sand, and convey the impression of a sterile and inhospitable coast, which is quite at variance with the indications of the country a a quarter or half a mile back from the shore.

The region about the head of Grand Traverse bay is mostly a level sandy plain, sufficiently elevated for drainage, but on the west and southwest of the head of the West Arm the country rises rapidly by one or two ascents into hills attaining an elevation of 300 or 400 feet. This elevation of the country is maintained most of the way to the Manistee river. The Monroe settlement lies in an elevated undulating expanse reaching south and east for six or eight miles. Toward the west of this the surface subsides, but remains dry to the head waters of the Becs Scies river.

The Peninsula is a gently hilly tongue of farming land. Similar features belong to the eastern shore of the east bay. Indeed, nearly the whole of the western part of Antrim county is made up of plains and gentle slopes, which sometimes reach an elevation of 200 feet, but toward the interior are ridges which attain a more considerable altitude.

The strictly low lands of the Grand Traverse region are scarcely worthy of mention. Occasionally a narrow belt of swamp borders a lake for a short distance, or spreads out in the vicinity of the mouth of a stream. Some low ground is observed about the head of the West Arm, and more about the head of the East Arm. The immediate vicinity of the upper waters of the Boardman river is somewhat marshy, as also some patches in the southeastern, middle and western parts of Grand Traverse county. The same may be said of the region about the head waters of the Bees Scies river, in Benzie county, as also the vicinity of Cedar river in Leelanaw. Some low ground occurs again about the south end of Elk and Round lakes, and on the borders of the streams in the interior and eastern parts of Antrim county.

Though the immediate shore, as seen from the lake, presents the appearance of a dune covered coast, we find very little sand blown toward the interior, except on the Sleeping Bear. Indeed, the beds of white material forming so striking a spectacle seen from the lake are more clay than sand; and I am not aware of any real dunes except in the region already indicated. The northern lakeward slope of Sleeping Bear Point consists of drifts of shining sand for a distance of two or three miles. The mound which constitutes "the Bear" is also clothed with drifted sand, though the vegetation growing upon it is evidence of a more coherent material beneath.

The scenery of the Grand Traverse region is subdued and soft—sometimes picturesque, always beautiful, in some instances exquisitely so. Viewed from some suitable eminence the landscape presents an undulating sea of verdure, one softly-rounded hill top succeeding another in the retreating view, the dimness of distance lending an ever increasing enchantment to the prospect. Frequently the introduction of water into the landscape gives it almost the perfection of enchantment. From the bluff on which the seminary of New Mission is situated the beholder has an exquisite view of Grand Traverse bay with its eastern and western arms dissolving in smoke in the dim distance, and the broad lake seen through the mouth of the bay sinking beneath the northern horizon. An emerald fringe of forest skirts the opposite shore ; the softened outlines of the Peninsula emerge from the misty embrace of the two arms of the bay, and all around the framework of this scene loom from the background the purple hilltops, looking perpetually down upon the picture.

From the foot of Pine lake a scene of surpassing loveliness presents itself. We land, perhaps, upon the wharf at the mouth of Pine river. Before us is a sandy slope on the top of which we discover the usual features of a new settlement. Beyond is the forest. It is a pleasant October morning, however, and we follow the well-beaten road through the fresh clearings which stretch out for two miles inland. We emerge from a screen of forest trees and find ourselves standing upon an elevated bluff overlooking as lovely a sheet of water as the sun ever shone upon. You feel almost a transport of delight in emerging so suddenly from the depths of the habitual forest into a prospect so vast, so gentle in its features, so delicate in its tints, and so glowing in the bright sunshine of a fair October morning. Far away to the southeast, for fifteen miles, stretches the placid smiling surface of the water, its white and pebbly shore chasing the contour of the hills in all its meandering sinuosities. The verdant ridges rise on every side from the shining shore line, and hold the lake in their enchanted embrace, while rounded hill-tops bubble up in rapid succession across the retiring landscape till hill and vale and sky, and green and purple and blue dissolve together in the blended hues of the distant horizon.

To one more of these views I cannot resist the temptation to allude. From an eminence about 400 feet high, two or three miles inland from Glen Arbor, on the northeast side of Glen lake, can be seen one of the most beautiful and varied landscapes to be witnessed in any country, and one which is well worthy the pencil of the artist. The view is toward the west, and it should be taken when the sky is clear and the atmosphere is pervaded by that softened haze which fuses the sharper angles of the landscape and throws over it a thin veil of inscrutable vagueness. From our hill summit we look down on the tops of the trees which cover the plain immediately fronting us. On the left is a portion of Glen lake, its nearer shore concealed by the forest, and the remoter one exposing a white and pebbly margin from which the verdant hills beyond rise hundreds of feet above the watery mirror in which their forms are so clearly fashioned. In front of us the green hills separate Glen lake from lake Michigan, and conceal from view the desert sand-fields of Sleeping Bear. Not completely, however, for the naked and glistening flanks of the northern slope stretch out to view beyond the forest-covered ridge, and embrace the placid harbor which struggles through the intercepting foliage, and blends with the boundless expanse of the great lake still beyond. Farther off in the midst of the water, rises the green outline of the South Manitou island, bearing on its head a glistening cap of sand. Still farther to the right rises the form of the North Manitou, which seems trying to hide itself behind the towering bluff of North Unity that guards the entrance to the harbor from the north. Two little lakes nestle in the rich woodland that spreads its verdure between us and the harbor, screening themselves like wood nymphs behind the thick foliage which half conceals their charms. It is doubtful whether a scene superior to this exists in the country.

V. SOIL.

The arenaceous element of the soil is generally strongly marked. At the same time the region on the west side of the bay is somewhat more sandy than that on the east. The soil of Grand Traverse and Benzie counties is more diversified. Nevertheless, patches of clayey soil are not unfrequent in Leelanaw county, and a well-mixed sandy loam is the dominant character of the soil on the hills. It seems, at first thought, somewhat surprising that the soil of the valleys should be less coherent than that on the slopes and summits of the hills. This disposition, however, is the natural result of the wasting of the hills by storms. These have worn away the more arenaceous materials and transported them to the lower levels, until the denudation of the hill summits has reached the beds of argillaceous materials with which all the hills of the region are intersected.

A considerable area about the head of the two arms of the bay is a sandy plain, the most of it sufficiently elevated for drainage. On the west of the bay the broken land reaches to the waters edge. On the south it is reached within two miles when a fine belt of adhesive loam extends for about five miles. This is succeeded by two or three miles of clayey soil less perfectly drained, after which we ascend to the beautiful plateau on which the Monroe settlement stands, clothed with a light loamy soil which extends southward with varying accessions of sandy material as far as the Manistee river. Eastward from the Monroe settlement the character of the soil continues to be a light loam, while toward the west and northwest it becomes more sandy and less perfectly drained. On the East of Silver lake is a region in which the argillaceous element decidedly predominates; while the country between New Sweden and Elk lake is favored with a well-drained calcareous loam, equal in fertility to any in the Grand Traverse region, and, from its having been longest settled, generally reputed to be somewhat superior to most parts of the country. This opinion, however, is an unwarranted disparagement of the country in general.

The soil on the east of Grand Traverse bay is a sandy calcareous loam of considerable uniformity, but yet, as on the west side, more sandy in the valleys than on the hill-tops. Benzie county presents diversities of soil similar to those of Grand Traverse county. The western border approximates Leelanaw county in its topography and soil. The southeastern part presents a continuation of the low sandy belt of the adjoining county.

In productiveness the soil of the Grand Traverse region is literally unsurpassed. The evidences of this will be seen when I come to treat of its farm crops and fruits. The proof of it is seen also in the astonishing magnitude of the forest tree which sustain themselves not merely upon the mould which has accumulated upon the surface, but strike their roots deep and

2

draw up stores of vegetable nutriment from the subsoil. The cause of the fertility of these soils is also apparent. Even the most sandy soil of Leelanaw county is unlike the sandy soils of other regions in its constitution. These sands have not been produced by the disintegration of sandstone strata, as is generally the case with sandy soils. There are no sandstone formations within the limits of the region. They are derived from the disintegration and decomposition of slightly arenaceous limestones. Pebbles of limestone are consequently more or less abundant in the soil—their abundance depending upon the proximity of the undisturbed formation. The continual solution of the calcareous matter of these limestone fragments furnishes a never-failing supply of lime to the soil, at the same time that it disengages additional amounts of sandy particles from their confinement in the limestone mass. These soils, therefore are naturally charged with the fertilizing constituent of plaster, which is lime—though it is probable that the sulphuric acid of common plaster exerts also some agency of which lime is incapable—and even this agency is supplied by the decomposing pyrites which the underlying rocks contribute to the soils of the region.

Aside from their habitual destitution of fertilizing constituents, arenaceous soils possess physical qualities favorable to productiveness. A sandy soil is always light. Atmospheric influences are allowed free access to the roots of vegetation, and to the soil constituents which need to be oxygenated for the purposes of agricultural utility. Even the tramping of men and animals fails to solidify them to the same extent as a clayey or even a loamy soil. A sandy soil is, besides, exempt from supersaturation with water; and yet it holds tenaciously water enough to answer the demands of vegetation. Through the free access of the atmosphere this water rapidly evaporates, thus surrounding the vegetable with vapor and affording the growing leaf the conditions most favorable to its health and expansion. Finally, a sandy soil is proved, by direct experiment, as well as by its promptness in bringing forward a crop, to be a more powerful absorbent of heat than a clayey soil, as well as slower to part with it. The sand is warm much sooner than the atmosphere and retains its warmth after the atmosphere has received its evening chill. Objection has been made to sandy soils, that their fertilizing constituents "leach out." Let us see. It is evident that whatever sinks into the earth, must go *in a state of solution*. No material particles can be supposed to descend, for we employ this very sand, in filters, to free water from its turbidity and sediments. Experiment proves that clean sand will even abstract some of the saltness from brine. But if the nutritive elements of the soil disappear in a state of solution in the water, there exists a union between them and the water which cannot be materially affected, under the actual conditions, until the water is again evaporated. In a period of dry weather, therefore, when sandy soils draw up by capillary attraction a supply of water from beneath, the same fertilizing constituents must return with it to the surface. Here the water, undergoing a rapid evaporation, deposits again the soluble ingredients which it had carried down at the time of the last rains. Thus it appears how nature has provided for the permanence of the fertilizing elements of the soil, and how drouths are a part of the agency employed by nature in preserving from waste the provision which she has made for the perennial nourishment of vegetation.

It appears, then, that the physical properties of sandy soils tend greatly to favor the development of vegetation, while, aside from the tendency to wash, it is only a deficiency in certain chemical constituents which has given sandy soils in general a bad reputation for being rapidly exhausted of their fertility. It is apparent, nevertheless, that sandy soils may exist not affected by such deficiency, and whose origin has been such that an adequate proportion of alkaline constituents has been supplied contemporaneously with the sand, and must continue to be supplied. The sandy soils of the Grand Traverse region are of this class. They possess, then, all the eminent recommendations dependent on the physical constitution of such soils, and all the chemical constituents which belong to strictly argillaceous or calcareous soils. Hence the secret of the enormous timber growth of the region, and its surprising agricultural productiveness.

Lest it should be objected that sandy soils, unsuited for farming purposes, do sometimes (though rarely, I think,) produce pines and hemlock of a large size, it may be well to remind the reader that the Coniferæ—embracing the pines, hemlock, cedars, firs and spruces—incorporate a large proportion of silicious matter in their constitution, and will flourish well on a soil more purely silicious than other (or gymnospermous) trees. Every one knows that the ashes of the Coniferæ are less desirable for potash manufacture than the ashes of the elm, ash, basswood, maple and beech. It is also notorious that a heavy forest of the latter class publishes a favorable account of the soil upon which they have been nourished.

VI. CLIMATE.

The climate of a region sustains a causal relation to its salubrity, its accessibility, and the character of its vegetable and animal productions. It is one of the most important considerations bearing upon its eligibility for business, settlement and homesteads. Climate depends principally upon three conditions—latitude, altitude above the sea, and relation to large bodies of land and water. The Grand Traverse region lies in about the same latitude as Nova Scotia, the middle of Maine, northern Vermont and New York, St. Paul in Minnesota, and Oregon City, Oregon. Its mean elevation above the sea being 800 feet, its mean temperature should be about two and one-third degrees lower than that of other places in the same latitude lying at the sea level. Or, since a mean annual difference of two and one-third degrees answers, in the temperate zone, to a difference of latitude of one degree and twenty-four minutes, the mean temperature of the year in the Grand Traverse region, in the mean latitude of 44° $52'$, should agreee with other places at the level of the sea in latitude 43° $28'$, which is about the latitude of Portland, Maine, Lockport, N. Y., and Milwaukie and Prairie du Chien, Wis.

I have had access to thermometrical observations, more or less complete taken at Traverse City (latitude 44° 46') by J. F. Grant, Esq.; at Northport (latitude 45° 08') by Rev. George N. Smith, and at Grand Traverse (latitude 44° 57') by Dr. H. R. Schetterly. An abstract of observations taken at Traverse City for six successive winters is given below:

Table I. Abstract of Meteorological Observations at Traverse City.

	HIGHEST.			LOWEST.			MEANS.			
	7am	1pm	7pm	7am	1pm	7pm	7am	1pm*	7pm	Day
1859–60.										
Dec.†	35	37	38	-6	9	8	17	24	20	20
Jan	46	52	45	-5	7	-7	19	30	22	24
Feb	42	49	45	-14	10	0	16	28	21	22
March	48	64	63	13	19	21	29	41	34	35
Winter	48	64	63	-14	7	-7	20	31	24	25
1860–1.										
Dec.†	35	37	35	-2	13	12	17	24	21	21
Jan	31	36	33	-11	10	1	17	24	20	20
Feb	44	59	44	-10	-12	-13	21	29	25	25
March‡	40	50	42	-8	-2	-6	21	30	23	25
Winter	44	59	44	-11	-12	-13	19	27	22	23
1861–2										
Dec	58	58	55	12	16	14	29	37	32	33
Jan	32	39	34	-9	-6	-1	14	25	20	20
Feb	34	44	33	-15	5	1	14	25	19	19
March	36	44	46	9	14	17	25	36	31	31
Winter	58	58	55	-15	5	-1	20	33	25	26
1862–3										
Dec	42	50	50	2	14	14	27	34	30	30
Jan	47	50	46	7	8	10	27	33	30	30
Feb	34	49	47	-10	8	3	18	29	24	24
March	38	46	40	15	19	9	25	32	27	28
Winter	47	46	50	-10	8	3	25	32	28	28
1863–4										
Dec	37	44	44	-2	21	10	24	31	27	27
Jan	34	45	39	-14	-4	-12	18	26	22	22
Feb	37	45	41	-14	-2	-9	22	29	24	25
March	39	54	44	-10	10	1	21	33	25	26
Winter	39	54	44	-14	-4	-12	21	30	24	25
1864–5										
Dec	40	46	37	-1	7	6	21	27	23	24
Jan	33	38	33	1	11	5	18	25	20	21
Feb	33	45	42	-6	5	-4	18	31	23	24
March										
Winter										
Five Winters	58	64	63	-15	-12	-13	21.0	30.6	24.6	25.4

*Observations were taken at noon during the winter of 1859–60.
†Observations began December 8th, 1859, and December 12th, 1860.
‡Ending with the 25th.

Table II. Comparison of Temperatures of the four Coldest Months at Various Places.

PLACES.	DECEMBER.				JANUARY.				FEBRUARY.				MARCH.			
	Mean max.	Mean min.	Extr. min.	Me'n tem.	Mean max.	Mean min.	Extr. min.	Mean tem.	Mean max.	Mean min.	Extr min.	Mean tem.	Mean max.	Mean min.	Extr min.	Mean temp.
Traverse City......	45.50	0.5	-6	25.67	43.33	-5.55	-14	22.83	48.50	-12.0	-15	23.16	52.0	2.6	-10	29.00
Manitowoc, Wis...	44.60	-3.6	-16	24.53	43.50	-10.75	-22	20.66	49.75	-9.0	-16	23.69	57.2	-0.4	-4	32.02
Hazlewood, Min...	38.50	-15.0	-28	13.09	38.80	-24.20	-32	7.20	45.83	-21.0	-28	12.60	50.8	-4.5	-15	27.11
Jt. Johnsbury, Vt..	42.75	-23.0	-34	17.43	43.50	-23.00	-40	17.75	45.00	-26.7	-31	13.71	51.4	-14.8	-27	28.83
Gardiner, Me......	43.25	-13.0	-25	20.35	40.60	-22.80	-32	17.50	45.60	-11.0	-18	18.68	49.8	-6.2	-10	29.28
Montreal, C. E.....	43.42	-14.62	-32	18.55	39.87	-22.00	-30	12.00	45.12	-20.5	-37	17.08	51.0	-4.7	-11	27.04
Portland, Or.......	57.00	22.0	22	41.51	52.00	20.00	20	38.11	54.00	28.0	28	39.86	58.0	32.0	22	42.47
Ann Arbor, Mich...	48.50	-0.65	-9	24.48	45.00	-11.00	-24	19.73	42.00	-12.0	-14	19.25	52.0	2.7	-10	29.60
Janesville, Wis....	47.00	-8.26	-20	22.94	43.00	-22.50	-29	11.50	44.60	-16.6	-24	18.96	57.2	-0.8	-6	28.84
Dubuque, Io........	48.40	-3.2	-16	24.28	47.20	-8.60	-20	20.80	46.60	-12.8	-20	21.33	62.4	9.2	1	35.65

Before proceeding to discuss the foregoing table it will be proper to present another one furnishing certain additional information in reference to the localities embraced in Table II. In order that comparisons of temperature instituted amongst different places may convey correct ideas, such comparisons ought to be made between corresponding years, and for long periods of time. Where the number of years embraced is few, and one of them happens to have been unusually mild or unusually severe, the effect upon the means is considerable. The periods of the observations upon which the results of Table II. were calculated are therefore given in the 4th column of Table III.

As altitude is also an important element in such comparisons, the altitudes of the places are given, as far as known, in the 3d column of Table III.

The latitudes of the same localities are given in the 2d column of Table III. The first seven places, it will be seen, are not far removed from the latitude of Traverse City. The remaining localities have been introduced into the discussion for the purpose of showing that places much further south possess a winter climate more severe than that of Traverse City.

Table III. Supplementary to Table II.

PLACES.	LATITUDE.	ALTITUDE AB'VE SEA LEV.	PERIOD OF OBSER.
Traverse City.............	44° 45'	525 ft.	1859–65
Manitowac................	44° 07'		1856–9
Hazlewood................	45° 00'		1954–9
St. Johnsbury.............	44° 25'	540 ft.	1854–9
Gardiner..................	44° 11'	75 ft.	1855–9
Montreal..................	45° 30'	118 ft.	1855–64
Portland..................	45° 24'	150 ft.	1859
Ann Arbor................	42° 16'	891 ft.	1854–7
Janesville................	42° 42'	768 ft.	1854–9
Dubuque	42° 30'	680 ft.	1854–9

The second and third columns of Table II. show for each place the mean of the December *maxima* and *minima* during the years covered by the observations—that is the mean of all the highest December observations for the several years, and the same of the lowest. The fourth column shows the lowest degree reached by the thermometer in December, during the whole period of observations for each place. The fifth column exhibits for each place the mean temperature of all the Decembers embraced in the period of the observations. The remaining columns of the Table give the same results for the months of January, February and March.

Of the localities lying nearly on the parallel of Traverse City, it will be observed that Manitowoc is located immediately on the western shore of Lake Michigan, and has Green Bay lying not over 35 miles to the north. It necessarily experiences therefore some modification of its winter climate from the influence of those large bodies of water. In this respect it seems even to be more favored than Milwaukie, 75 miles further south, which has colder winters—the difference, perhaps, being the measure of the influence of Green Bay upon the winter climate of Manitowoc. Portland, Oregon, is under the influence of the Pacific ocean, as the observations show. Hazlewood, Min., Montreal and St. Johnsbury are situated inland, and may be taken as fairly representing the continental temperature on their parallels, as unmodified by large bodies of water.

The adaptation of a winter climate to the safe wintering of fruit trees and farming crops is not indicated by the mean temperature of the winter, nor by the mean temperature of the several months. Nevertheless, when this comparison is made, we perceive that the climate of Traverse City is milder than that of any other locality given in the table—Portland, Oregon, of course, excepted. In the month of December Manitowoc is over one degree colder; Hazlewood, 12½°; St. Johnsbury, 8⅛°; Gardiner, 5¼°; Montreal, 7°; Ann Arbor, 1¼°; Janesville, 2¾°; Dubuque, 1°.

In the month of January Manitowoc is 2¼° colder than Traverse City; Hazlewood, 15⅜° colder; St. Johnsbury, 5°; Gardiner, 5⅛°; Montreal, 10¾°; Ann Arbor, 3°; Janesville, 11⅛°; Dubuque, 2°.

In the month of February Manitowoc is half a degree warmer than Traverse City; Hazlewood, 10¼° colder; St. Johnsbury, 9½°; Gardiner, 4½°; Montreal, 6°; Ann Arbor, 4°; Janesville, 4°; Dubuque, 2°.

In the month of March the mean of the more southern localities begins to feel the influence of occasional warm southerly and southwesterly winds, while Traverse City is still environed by the winter temperatures imprisoned in the ice of the bay. It is the extremes of winter temperature which produce such frequent destruction of the more delicate varieties of fruit trees. The table furnishes the mean *minima* of the several places for the cold months of the year. In December the mean *minimum* of Manitowoc is 4° lower than at Traverse City; of Hazlewood, 15½° lower; of St. Johnsbury, 23½° ; of Gardiner, 13½°; of Montreal, 15°; of Ann Arbor, 1°; of Janesville, 8¾°; of Dubuque, 3¾°.

In January the mean *minimum* of Manitowoc is 5¼° below that of Traverse City; of Hazlewood, 18½°; of St. Johnsbury, 17½°; of Gardiner, 17¼°; of Montreal, 16¼°; of Ann Arbor, 5¼°; of Janesville, 17°; of Dubuque, 3°.

In February the mean *minimum* of Manitowoc for the years compared is 3° higher higher than at Traverse City; of Hazlewood, 9° lower; of St. Johnsbury, 14¾° lower; of Gardiner, 1° higher; of Montreal, 8½° lower; of Ann Arbor, the same; of Janesville, 12½° lower; of Dubuque, ¾° lower.

The mean *minimum* for March is lower for every one of the places compared with Traverse City, except Portland, Oregon.

The favorable character of the winter climate of Traverse City is placed in a still stronger light if we compare the *extreme minima* for a series of years. The mean *minimum* may be of moderate severity, while on one or two occasions in the course of the winter, or still more likely within a range of five or six years, the mercury may sink to the damaging limit. The *extreme minimum* of Manitowoc compared with that of Traverse City is seen to be, in the month of December, 10° lower; of Hazlewood, 22°; of St. Johnsbury, 28°; of Gardiner, 19°; of Montreal, 26°; of Ann Arbor, 3°; of Janesville, 14°; of Dubuque, 10°.

In January. the *extreme minimum* of Manitowoc is 8° lower than at Traverse City; of Hazlewood, 18°; of St. Johnsbury, 26°; of Gardiner, 18°; of Montreal, 16°; of Ann Arbor, 10°; of Janesville, 15°; of Dubuque, 6°.

In February, the *extreme minimum* of Manitowoc is 1° lower than of Traverse City; of Hazlewood, 13°; of St. Johnsbury, 16°; of Gardiner, 3°; of Montreal, 22°; of Ann Arbor, 1° higher; of Janesville, 9° lower; of Dubuque, 5° lower.

It thus appears that under every point of view the winter climate of Traverse City is materially milder than that of other places in the same latitude either east or west. It is materially milder than that of places two and a half degrees further south. The *minimum* range of the thermometer being but 15° below zero, it does not reach the point at which peach trees are injured; and in this respect the winter climate compares favorably with that of middle Ohio, Indiana and Illinois. Indeed, the winter extremes for ten years past, during which peach trees have been growing in the Grand Traverse region, have been less than at Cincinnati or St. Louis, or even Memphis, in Tennessee. During the memorable "cold spell" of New Year's 1864, the thermometer is reported to have sunk at Milwaukie and Janesville, Wis. to 40° below zero; at Chicago, to 29° below; at Kalamazoo, Mich., to 20° below; at St. Louis, to 24° below; and at Memphis, Tenn., to 16° below. The following figures exhibit the movement of the mercury at Northport and Traverse City during the same interval:

Table IV.   *Observations during the cold cycle of* 1863-4.

1863-4.	NORTHPORT.			TRAVERSE CITY.		
	7 A. M.	2 P. M.	10 P. M.	7 A. M.	1 P. M.	7 P. M.
Dec. 31.	22	28	18	20	29	28
Jan. 1.	0	–8	–14	8	–2	–12
,, 2.	–14	–6	–3	–14	–4	–3
,, 3.	–3	2	8	8	10	3
,, 4.	4	11	7½	1	17	8
,, 5.	11	16	8	8	19	12

This cycle of cold weather, which extended over the entire northwest and destroyed or damaged fruit trees in every northwestern State, caused no damage whatever in the Grand Traverse region.

Another characteristic of the winter of this region is its comparative uniformity of temperature. The mercury neither rises as high nor sinks as low as in other regions along the same parallel of latitude.

Other comparisons are no less surprising than those which have just been made. Autumnal frosts are postponed to a remarkably late period. Unlike other regions, frost seldom appears till the mercury actually reaches 32°. The first killing frosts ordinarily occur throughout the region, between the middle and end of October. Sometimes they are delayed till late in November. They occur at Traverse City and southward from there somewhat earlier than at Northport, Glen Arbor and Frankfort. The first killing frost this year at Traverse City was a slight one, October 13th, but it did not reach Northport. Tomatoes and other tender vegetables were still growing thriftily at Northport and Pine River, and even at the head of Little Traverse bay, when I visited those places, Oct. 27th and 28th. On the night of the 28th, however, the thermometer sank to the freezing point, and injured vegetation generally throughout the region. On the 5th of November, it froze again. At the same time the mercury sank to 24° at Ann Arbor, and to zero at Bangor, in Maine. Nevertheless, when I left the region on the 8th of November, the leaves of apple and peach trees were still perfectly green, while those of the forest were partially changed and beginning to fall. On reaching the southern part of the State, vegetation presented already the appearance of mid-winter.

Autumnal frosts occur only after days of very threatening severity. I observed that when, during the day, the thermometer rises as high as 40°, it is seldom crowded down to the freezing point the following night. At Ann Arbor we often get frost after the thermometer has been at 60° during the previous day.

Snow falls in November or December, before the ground has been materially frozen, and lies without thawing till the following April. It accumulates to the depth of two or three feet, and sometimes, in certain localities, to a greater depth. Its disappearance is postponed till abouth the 10th of April, when the danger of severe frost is generally passed. The ground consequently escapes freezing throughout the entire winter, so that root crops may be left out without damage. Potatoes are thus, frequently, wintered in the ground without digging. It always happens that the few remaining in the soil after the crop has been gathered, vegetate in the following spring, and produce a spontaneous crop. Thus they propagate themselves from year to year, so that the Irish potato has become a naturalized weed, growing in corn fields and wheat' fields, and sometimes in uncultivated fields, and by the road side. I saw potatoes growing in places where I was informed no seed had been planted for ten years.

The same preservative effects of snow are witnessed in other crops, and in the bulbs, tubers and roots of ornamental plants. The Dahlia blooms till the last of October, and after this the tubers may be left in the soil till the following spring, when, not long after the disappearance of the snow, they send up fresh shoots. Delicate green-house roses stand out with the same impunity as in Alabama and Louisiana. Mrs. Judge Fowler, of Mapleton, on the Peninsula, informed me that she had in her garden forty varieties of delicate roses, which stand out every winter.

Wheat, of course, is never in danger of winter-killing in a region thus exempt from extremes of cold, and thus clothed during the entire winter with a thick mantle of snow.

The presence of snow till the middle of April preserves vegetation from the stimulating influence of occasional warm days, and the buds of fruit trees consequently remain dormant till the danger of severe frost is passed. When the snow finally disappears, the soil is in a condition to receive immediately the genial influence of sunshine and atmospheric action. The disagreeable period of mud caused by the slow escape of frost from the soil is unknown. The breaking up of the ice in the bay exposes the entire region to the equalizing influence of large bodies of water, and the region is thus nearly as exempt from the destructive effects of late vernal frosts, as from those of late autumnal ones. No damaging frost is liable to occur later than the middle of May, which is about the period of latest frosts in northern and middle Ohio.

The mean temperatures of the four winter months at Grand Traverse, for five years, are as follows :

December... 25°.2
January.. 23°.2
February... 23°.0
March .... 29°.0

The following are the means of four months of the year 1860 at Northport:

January.. 22°.48
February... 22°.91
March .... 33°.91
April .... 40°.33

The temperature of summer is as remarkable for its moderation and uniformity as that of winter. I have not had the opportunity to examine any record of thermometrical observations made during the summer, but the summer climate is admitted to be exempt from extremes and sudden changes. Yet the mean temperature is sufficiently high to mature peaches, tomatoes, tobacco and the like.

The facts which I have disclosed above touching the winter climate of the Grand Traverse region are well calculated to excite surprise; but I think no one can question the figures. A moment's reflection, moreover, will reveal the reason for the peculiarities of the climate of this portion of the State. The Grand Traverse region, like the peninsula of Florida, Sweden and the British islands, is subjected to the equalizing influences of large bodies of water. Lake Michigan borders the whole western slope of the State. In the region under consideration the body of water is greatly augmented by the bay which reaches its two arms thirty-four miles into the interior. Moreover, the triangle forming Leelanaw county is embraced by two large, bodies of water, and enjoys a situation unlike that of any other portion of the northwestern States. Our cold winds generally proceed from the southwest or west. Passing over the open water of Lake Michigan sixty miles in width, the temperature of which never sinks below 32°, it is impossible to avoid ab-

stracting a considerable amount of heat, so that when these
cold westerly winds strike the Michigan shores of the lake, the
severity of the winter gales is materially mitigated. Moreover,
the severest and most destructive winter gales proceed from
the southwest, and the trend of the lake is such that these
winds, on striking the Grand Traverse shore have traveled
over more water than southwest winds striking the Michigan
shore in Ottawa, Van Buren and St. Joseph counties.

But the thermometer on some occasions sinks to a *minimum*
with an easterly or even a southeasterly wind—as in February,
1857, when it sank twenty-four degrees below zero at Ann Arbor with an easterly wind and a cloudy sky. Before such
winds the eastern shore of lake Michigan in St. Joseph county
and northward experiences no protection from the proximity of
a large body of water. In the Grand Traverse region, on the
contrary, the diameter of the peninsula is so much diminished
that easterly winds retain the softening influence exerted by
the waters of lake Huron. Moreover, the whole of Leelanaw
county enjoys nearly as complete protection from easterly
as from westerly winds. It is almost impossible for a gale
from any direction to bring into Leelanaw county a temperature of eighteen or twenty degrees below zero, the point at
which the limbs of peach trees are liable to be killed.

No observations on the other elements of climate have been
brought under my observation. It is obvious, however, that a
region so environed by water must possess an atmosphere of
sufficient humidity to offer a guarantee against habitual
drouths. I am informed that no severe drouth has ever been
experienced before the summer of 1864, when the whole northwest was parched to an unprecedented extent.

VII. SALUBRITY.

, A region possessing such a climate, and such physical features as have been described above, can scarcely offer any other
than favorable sanitary conditions. Accordingly, I was everywhere assured by the inhabitants of the region that diseases
are almost unknown. I heard of a few cases of typhoid fever
in the neighborhood of Glen Arbor, and a few cases of dysen-

tery about Leland and in Antrim county. Bilious diseases are foreign to the country. No ague was ever known to be indigenous to the region. On the contrary, many chronic cases of suffering from malarious influence have been relieved and cured by a residence in the region. The uniformity of the temperature and the purity of the air and water are also favorable in pulmonary diseases; and I learned of some rheumatic affections that had been cured by a few months residence.

VIII. TIMBER AND NATIVE PLANTS.

Passing from a survey of the physical features of the Grand Traverse region, I proceed to offer a brief account of its natural history. Generally speaking the region is covered by a magnificent growth of hardwood timber. The exceptions to this statement are few and unimportant. By far the most abundant species is the sugar maple (*Acer saccharinum*). This distributed generally throughout the region on both sides of the bay. It bears, however, a larger ratio to the whole forest on the west side. Mingled with this are the beech (*Fagus sylvatica*), white or American elm (*Ulmus Americana*), and hemlock (*Abies Canadensis*). The beech, as might be expected, is more abundant on the more coherent soils of the east side of the bay and in Grand Traverse county. The hemlock is pretty generally scattered through the forest of Leelanaw, Grand Traverse and Benzie counties, forming on an average about one fifteenth, or less, of the forest growth. It occurs less frequently in Antrim county. In certain situations where the soil is most retentive we encounter patches of forest diversified with the black ash (*Fraxinus sambucifolia*), while the arbor vitæ, or western "white cedar" (*Thuja occidentalis*) holds joint possession with the balsam fir (*Abies balsamea*), in some moist and wet lands, and the tamarack (*Larix Americana*) sometimes crowds itself into the company of the other denizens of the occasional swamps. The white pine (*Pinus strobus*) is very partially distributed. Some majestic specimens—individuals of which attain a diameter of nearly five feet—may be seen in the south part of Leelanaw county on the east of Cedar run, where some wasteful settlers are engaged in felling them in winrows and wickedly burning them. A valuable belt of white pine lies in the southeastern part of Benzie county on the upper waters of the Becs Scies river, and another on the upper waters of Boardman river in Grand Traverse county, whence the logs are floated to Traverse City and worked up in the mill of Hannah, Lay & Co., which, according to the "Statistics of Michigan, 1864," produced in 1863 10,200,000 feet of pine lumber, worth $112,000. The product of the present year is said to be twelve million feet. Another pinery exists in the interior of Antrim county on the tributaries of Grass lake. The logs from here are worked up in the mill of Dexter and Noble at Elk Rapids, which, according to the authority above quoted, produced, in 1863, 4,000,000 feet, valued at $10,000—a valuation which would seem to be erroneous. The product of the present year is probably nine millions of feet. The county of Leelanaw is also reported to have produced in 1863, 395,000 feet of lumber (probably but little of it pine lumber), valued at $91,500.*

Occasionally, as on the Peninsula, I noticed the Norway pine (*Pinus resinosa*) in company with the white pine, sparingly dispersed through the forest.

The oak is not regularly distributed; but in certain regions it constitutes an important feature. I observed the red and white oaks (*Quercus rubra* and *alba*) growing abundantly on the sandy plains about the head of the two arms of Grand Traverse bay. I noticed the red oak growing also at Elk Rapids, and both oaks on the Peninsula. A grove of white oaks interspersed with black oaks, occupying 200 acres, flourishes on the north side of Round lake. I saw them also on the shores of Crystal lake in Grand Traverse county ; on the ridges back of Glen lake ; between Carp lake and Sutton's bay and in many other places. The trembling aspen or poplar (*Populus grandidentata*) is quite frequent about the borders of clearings—especially on the Peninsula—while the balsam popular (*Pop-*

---

*These statistics, taken from the work referred to, disclose some curious discrepancies. Following the figures, one dollar buys 4 feet of lumber in Lenawee county, 90 feet at Traverse City, and 400 feet at Elk Rapids.

ulus balsamifera) is also occasionally seen in all parts of the region. I was greatly interested to notice this tree struggling for an existence on the bleak and sterile plateau of the Sleeping Bear Point. Its gnarled and miserably dwarfed condition proclaimed the nature of the conflict it had endured; and a wonder arises why a tree so ill adapted to the situation should attempt to establish itself where nothing else can maintain an existence.

The yellow birch (*Betula excelsa*) is a frequent denizen of the forest, and sometimes grows to an extraordinary size. A specimen seen in Antrim county measured eleven feet and four inches in circumference four feet above the ground. The false white birch (*Betula populifolia*) is also frequently encountered. The black cherry (*Cerasus serotina*) is not unfrequent, and sometimes becomes a troublesome intruder on the borders of clearings. The soft maple (*Acer rubrum*) occurs sparingly about Antrim and probably in other localities.

This primitive forest presents to the eye of the traveller a scene of wonderful majesty, magnificence and interest. The towering hemlocks with their straight cylindrical trunks often three, four, or nearly five feet in diameter expand their crown of dark green spray at the summit, while the majestic maple, beech and elm lift their heads to an equal altitude, and mingle their paler and brighter foliage with that of the sombre evergreen. The undergrowth is scant, consisting of the striped maple (*Acer Pennsylvanicum*) and witch hazel (*Hamamelis Virginica*) with patches of dwarf yew or ground hemlock (*Taxus Canadensis*)—the latter of which is much more abundant on the west side of the bay. The forest is therefore an endless colonnade of majestic pillars; and, but for the prostrate forms of the fallen patriarchs of the wood, a vehicle could be driven through the unbroken forest from one end of the region to the other.

Some of the low grounds in the vicinity of Glen Arbor are covered with the cranberry vine (*Oxycoccus macrocarpus*) and the half shaded borders of the lakes are often clothed with the evergreen bearberry (*Arctostaphylos uva-ursi.* In most half-open situations the blackberry (*Rubus villosus*) and red

3

raspberry (*Rubus strigosus*) flourish luxuriantly and afford an inexhaustible supply of fruit. The raspberry sometimes produces two crops in the season. I saw at several places, ripe fruit, green fruit and flowers existing together in the latter part of October, upon canes of the current year's growth. The huckleberry (*Gaylussacia resinosa*) is also abundant in the sandy clearings about the head of the bay. The native strawberry of the region is *Fragaria Virginiana*, and it may be seen in fruit from June to October.

A singular and interesting assemblage of trees and shrubs covers the Sleeping Bear proper, and by its dark green foliage strongly isolates this pile of sand from the shining desert around it. The only trees upon the mound, besides some dwarfish balm of Gileads, are the balsm fir and white cedar. A stump of one of the latter, cut by the operators of the Lake Survey, measured two feet in diameter. The shrubs consist of the trailing red cedar (*Juniperus Sabinæ*), choke cherry (*Cerasus Virginiana*), dogwood (*Cornus florida*), snowberry (*Symphoricarpus occidentalis*), bearberry, wild rose, (*Rosa blanda ?*) and frost grape (*Vitis cordifolia*). The fruit of the latter, as well as that of the choke cherry was found to be quite palatable, though the leaves of the grape seemed to be uninjured by the frost, as late as the 24th of October. It is quite certain that the southern side of this lofty mound of sand would ripen the Isabella or Catawba grape with complete success.

It will be remarked that the vegetation of the region does not present a northern aspect. The northern white birch (*Betula papyracea*) is wanting, and the fir and spruce are but very feebly represented. The native strawberry is the Virginian species, and the pendent lichens, so marked a feature of the lake shore forest on the opposite side of the Peninsula, in the same latitude, are entirely unknown.

IX. ANIMALS.

It is not intended under this head to attempt an enumeration of all the animals of even a single class ; but only to bring

together a few statements of greater or less economical interest. The most conspicuous mammalian quadrupeds are the
black bear (*Ursus Americanus*) the Virginian deer (*Cervus
Virginianus*), the red fox (*Vulpes fulvus*), the otter (*Lutra
Canadensis*) and various squirrels—among wich I noticed the
black variety quite frequently. The bear is nearly confined to
the remote portions of Benzie, Grand Traverse and Kalkasca
counties. Of birds, the ruffed grouse or partridge (*Bonasa
umbellus*) is the most important, and affords a valuable article
of food. The quail (*Ortyx Virginianus*) has recently been
observed in considerable numbers. Of fishes, the usual lake
species occur in the bay, but not in such numbers to render
fishing a business of much importance. The speckled trout
(*Salmo fontinalis*) occurs plentifully in all the streams of the
region, and in many of the small lakes. Among insects, the
mosquito (*Culex pipiens*) is one of the most conspicuous ; but
it is worthy of mention that its disappearance for the season
occurs as early as July or August. Indeed, the mosquito is
much less troublesome in the Grand Traverse region than at
Chicago.

Another insect likely to become important from its ravages
committed upon vegetation, is the canker worm moth (*Anisopteryx pometaria*), and I think it worth while to present some
account of its nature and habits.

In the latter part of May or early in June the young leaves
of the maple, elm and some other trees are seen to be infested
by myriads of " worms" (so-called), which are the larves of
an insect belonging to the tribe of moths. These larves are
furnished with legs near the extremities only, and consequently
move with a measured motion which has caused them to receive the name of "span worms" or "measure worms." The
fullgrown larves attain the length of an inch or more. A great
difference of color is observable among the worms of different
ages. The young are generally of a blackish or dusky brown
color, with a yellowish stripe on each side of the body, which
is also whitish on the under side. When fully grown they become paler on the back, and a black line appears above the

yellowish one on the sides. In some a greenish-yellow color prevails; in others, a clay color, with dark lines or spots variously distributed on the sides and back. They commence their depredations by eating small holes in the leaves, which they continue to enlarge till, in some cases, the entire pulp of the leaf disappears, leaving little more than the mid-rib and veins. They attain full growth in June, when about four weeks old, and then begin to quit the trees. Some creep down the trunk; but great numbers let themselves down by their threads from the branches. After reaching the ground, they immediately burrow in the earth to the depth of from two to six inches. Here, by repeatedly turning and fastening the loose particles of earth by a few silken threads, they form little cells, and in twenty-four hours have changed to chrysalids. The chrysalis is about half an inch in length, of a light brown color, largest in the females. In the autum, after the appearance of frosts, the perfect insects begin to make their appearance from the crysalids. They appear at various times when the weather is sufficiently mild, from October to March. I saw countless numbers of the males fluttering over the surface of a light fall of snow, when riding through the woods early in November. I had seen them previously, during a cold rain, in October.

The females of the perfect insect are destitute of wings, and hence do not wander far from their place of birth. The males have a body about an inch and a half in length, and measure about an inch across the wings. The wings are rather large, thin, and covered with a grayish silky down. The hind wings are paler in color than the others. The females, on their appearance, whether in autumn, winter or spring, make their way toward the nearest trees, and creep slowly up their trunks. A few days afterwards the males may be seen fluttering about them, when, after pairing, the females deposit their eggs in clusters of from sixty to one hundred, upon the branches of the trees. The eggs are glued to each other, and to the bark, by a grayish varnish impervious to water. Their task having been accomplished, the insects soon die. About the middle or last of May the eggs are hatched, and a brood of worms appears upon the young foliage of the forest and fruit trees, which furnish these larves with their appropriate food.

Feeling uncertain whether the canker worm moth of the Grand Traverse region is the same as the common species of New England, I forwarded some specimens to A. S. Packard, Jr., M. D., of the Boston Society of Natural History, who informed me that it is not the common species, but probably another one sparingly distributed over New England, and known by the name of *Anisopteryx pometaria*, Harris.

It is a fact of scientific, as well as agricultural importance, that this species exists in such abundance in a region so isolated and so remote from Boston, which seems to have been the metropolis of the canker worms.

I was informed that this insect caused considerable damage, the past season, to apple orchards in the vicinity of Elk Rapids. As its depredations are likely to increase for some time, instead of diminishing, it will be well to adopt some measures of restraint. These, fortunately, are not difficult or expensive. The most effectual means are based upon the circumstance that the females are wingless, and are hence obliged to crawl up the trunks of the trees to deposit their eggs. Any means which will arrest the progress of the female will protect the tree. The editor of the *American Agriculturist* recommends covering the ground about the base of the tree with ashes. In New England, tar is frequently employed, by smearing a narrow belt around the tree trunk, or by closely girding the tree with a band of clay, paper, tin or cloth that has been covered with tar. In this case, care must be taken that no room is left between the tree and the girdle, through which the insects can ascend. Various other means have been recommended, for information on which the reader may consult Harris' "Insects Injurious to Vegetation," Boston, 1862. It is obvious that whatever preventives are employed, must be resorted to before the chrysalids are changed. There are some reasons for believing this species to come out of the ground, principally in the autum, which is contrary to the habit of the common species (*Anisopteryx vernata*, Harris). The remedies must, therefore, be applied in October and November, and renewed again whenever the weather becomes mild enough to stimulate the insects to activity.

Our warfare upon them need not cease after the appearance of the "worm" upon the trees in May. They may be destroyed by sprinkling on them finely pulverized quick-lime mixed with ashes, at times when the leaves are wet. Whale oil soap suds sprinkled on them will also kill them, as it does most other insects. These remedies may be employed where the number or size of the trees to be protected is not too great.

X. GEOLOGY.

1. GENERAL GEOLOGY.—The general geological structure of the Lower Peninsula of Michigan may be learned by reference to my "Report of the Progress of the Survey of Michigan for 1859–60." The Grand Traverse Region, like most parts of the Peninsula, is covered with drift materials of such depth that exposures of the underlying strata are quite rare. In order, therefore to investigate the geology of the region, it is necessary to extend our observations considerably beyond its limits to other regions where the strata underlying this region rise to the surface and present exposures. With this view, my personal examinations on the present survey were begun at the head of Little Traverse bay, extending thence southward into Benzie county. The results embodied in the following pages, however, are drawn, to a large extent, from observations made in all portions of the Peninsula during the continuance of the public geological survey. Confirmations of the conclusions arrived at from the examination of outcrops beyond the limits of this region, have been sought in the soundings of the United States Lake Survey, and in the constituents of the drift materials of different parts of the region.

The information which I have been able to obtain from all sources shows that the strata underlying the northwestern slope of the Lower Peninsula of the State have a gentle dip southeastwardly, toward the middle of the Peninsula. This dip of the strata is preserved in traveling northwest across Lake Michigan into the Upper Peninsula, and was undoubtedly communicated to the strata at the time of some eruption and uplift occurring in the Upper Peninsula—perhaps at the period of the eruption of the trap of Keewenaw Point. We are not aware of the occurrence of any geological disturbances of later date than this eruption; and there are, therefore, some grounds for presuming that the disturbance which tilted the Lake Superior sandstone, tilted also the other strata ·of the Lower Silurian and even of the Upper Silurian and Devonian systems—extending its effects as far as the Lower Peninsula. Aside, however, from the general dip of the strata within the region under consideration, there are no considerable evidences of disturbance. The rocks along the south shore of Little Traverse bay present a series of broad, wave-like undulations; and, on the east shore of Grand Traverse bay are seen two cases of abrupt dislocation, accompanied by a downthrow of three or four feet.

The following table presents a view of the formations which will be brought under consideration in the present report:

Quaternary System . . . . { Lignite. / Drift.
Lower Carb. System . . . . . . Marshall Group.
Devonian System . . . . { Huron Group . . { Light Shales. / Black Shales. / Hamilton Limestones. / Corniferous Limestone.
Upper Silurian System . . . . Onondaga Salt Group.

2. SALINA GROUP, OR ONONDAGA SALT GROUP.—This series of argillaceous limestones, shales, sandstones and gypsum underlies the entire Peninsula, outcropping at the base of Mackinac, Bois Blanc, Round and Little St. Martin's islands, and on the main land west of Mackinac. Passing thence under the Peninsula, it reappears in Monroe county, Michigan, and Sandusky bay, Ohio. On the west of Lake Michigan it outcrops, to a limited extent, near Milwaukie, and on the east, at Galt and that vicinity, in Canada West. It may come to the surface in some portions of the region between Little Traverse bay and the Straits of Mackinac—a region not yet geologically explored. It furnishes the gypsum of the Grand River, in Canada, of Sandusky, Ohio, and of Onondaga county, New York. It abounds in the same deposit at Little Point au

Chene, west of Mackinac. It is also the source of the supply
of brine to the salt wells of New York, and to those at St.
Clair and Port Austin in this State. It underlies the Grand
Traverse region, but the places of outcrop of the formation are
far beyond its limits.

3. CORNIFEROUS LIMESTONE.—The elevated limestone region south of the Straits to Little Traverse bay belongs to this
formation. It thence dips under the Peninsula and reappears
in Monroe county and the northern portions of Ohio and Indiana. Its western outcrop is in the bed of lake Michigan, and
its eastern is under lake Huron. It is the most conspicuous
and important limestone formation in the the Lower Peninsula,
affording its principal supply of quicklime, and furnishing in
places a building stone of considerable value. It is often sat-
urated to a remarkable extent with petroleum and black bituminous matter, causing it to be generally regarded by the un-
informed as an oil producing rock—an error which, in numerous instances, has only been discovered and admitted, after
many thousand dollars had been wasted, in contempt of scientific authority. The formation does not outcrop within the
limits of the Grand Traverse region as I have defined it.

4. LIMESTONES OF THE HAMILTON GROUP.—As the present survey has thrown much additional light upon the geology of this formation, I shall proceed to give some account of
the rocks belonging to it, at each outcrop within the limits of
our region.

Near the head of Little Traverse bay, south of the point—
E side sec. 23 T. 35 N. 5 W. (Locality 855 of the Geological
Survey) the following section is seen:

Section at 855.

E. Limestone, light, argillo-calcareous, in beds from 1 to
8 inches thick.................................. 4 ft.
D. Limestone, rather thick-bedded, somewhat porous,
with an uneven fracture, and containing crystals of
calcareous spar.................................. 4 ft.
C. Limestone, thin-bedded, light colored, argillo-calcareous, breaking into small angular fragments, contains corals .... 6 ft.
B. Limestone, much broken, in beds from one to two feet
thick... 6 ft.

A. Limestone, pale buff, argillaceous, banded, containing
fossils... 2 ft.
The dip at this place is about five degrees toward the southwest.

About twenty rods west of this locality is a more considerable exposure (856) extending along the beach for three-fourths
of a mile and forming an escarpment which, at some points attains an elevation of thirty-five feet. The stratification is mostly
horizontal, presenting slight undulations within distances of ten
or twelve feet. Toward the western part of the exposure the
cliff partially subsides. The trend of the coast is about southwest; and when, near the mouth of Bear creek, the trend
changes to west, the escarpment strikes inland by continuing
toward the southwest. The following succession of strata may
be made out at this outcrop:

Section of Stromatopora Beds (856.)

D. Limestone, pale buff, very massive, breaking in regular blocks, somewhat arenaceous, inseparable
from the following, except in its structure; contains the same fossils........................ 12 ft.
C. Limestone, pale buff, massive, brecciated in places,
vesicular; falling down in huge blocks. Becomes
more regular westward. It has a rude concretionary structure from the abundance of *Stromatopora* (with large cells)..................... 20 ft.
B. Limestone, thin-bedded below, thicker above, broken, with a 10 inch band of dark soil (bituminous) at top and thinner ones below. Contains
Atrypa reticularis, Favosites Alpenensis, &c.... 10 ft.
A. Talus or sloping beach of fragments............ 4 ft.

The *Stromatopora* at this exposure is by far the most abundant fossil, constituting the principal mass of No. C, and ren-
dering it a veritable coral reef. Fine specimens of this coral
can be collected in any quantity, even to a shipload. The in-
dividual masses are sometimes of an elongate-ovoid form, from
two to four feet in length, with the concentric plates arching
across the mass. These forms are found both erect and prostrate. At other times they are more spherical, attaining a diameter of four or five feet, and giving the rock or reef a confused concretionary structure. From the abundance of this
coral I shall designate the strata in which it abounds as the
"Stromatopora Beds."

The stratum No. C embraces also the following fossils: *Atrypa reticularis, Merista, Spirigera concentrica, Zaphrentis, Heliophyllum, Favosites, Cystiphyllum, Stromatopora* (smaller cells), *Cladopora, Conocardium, Tropidoleptus carinatus,* and other species.*

In the fields half a mile back from the bay I picked up *Strophomena nacrea, Acervularia Davidsoni, Atrypa reticularis* and *Favosites Alpenensis.*

At the saw-mill on Bear Creek, half a mile above its mouth is an outcrop of dark brown, very bituminous limestone, abounding in *Heliophpllum, Favosites Alpenensis, Atrypa reticularis.* It is clearly the same bed as the dark bituminous limestone about Thunder Bay, where, as well as here, its place seems to be above the Stromatopora Beds.

About a mile southwest from Bear creek we find a considerable exposure of a very different sort of rocks. This is on the S. E. ¼ Sec. 1 T. 34 N. 6 W. (Locality 857). The rocks are seen to have a dip varying from 5 to 15 degrees, and extending along the beach for 360 feet, with a mean dip of about 8 degrees, which would give a thickness of 53 feet to the whole exposure. The following is a statement of the stratification:

Section of Tropidoleptus Beds (857.)

E. Limestone, argillaceous, sub-crystalline, the thinner layers shaly, terminated by a few inches of black shale.................................... 14 ft.

D. Limestone, very dark chocolate colored, argillaceous, compact, much broken...................... 3 ft.

C. Limestone, very dark, bituminous, in beds from six inches to one foot thick, shaly or subcrystalline. 12 ft.

B. Limestone, dark brown, argillaceous, uneven bedded, breaking with a ragged uneven fracture.... 5 ft.

A. Limestone, dark, compact, argillo-calcareous, breaking with a smooth conchoidal fracture, much shattered. .... 1 ft.

The following fossils occur in the member A: *Favosites Alpenensis, Acervularia Davidsoni, Phillipsastræa Verneuili,* a

---

*Little has been done in the paleontology of this region but to make identifications on the spot, as far as I was able. The lists of fossils given may seem to possess but little practical importance. They possess, however, a scientific importance, and, indirectly, a practical importance; and I feel assured there will be not a few readers of this Report who will scan these lists and geological sections with interest.

branching Cyathophylloid, *Fenestella* (small fenestrules) *Fenestella* (large fenestrules) *Serpula*, *Strophomena nacrea*, *Strophomena* (large), *Tropidoleptus carinatus*, *Spirifer mucronatus*, *Spirifer sp?*

The following occur in the three upper members: *Favosites Alpenensis*, *Zaphrentis*, *Acervularia Davidsoni*, *Chonetes* (small, flat,) *Chonetes* (winged, large,) *Atrypa reticularis*, *Terebratula Romingeri*, *Nuculites*.

The strata at this locality seem to be the same as at the saw-mill on Bear creek, and their position is apparently above the Stromatopora beds at 856.

Proceeding about 100 rods along the shore we reach 858, a bluff 50 feet high, mostly covered by clay, pebbles and and sand. Some rocks containing *Acervularia Davidsoni* outcrop near the base of the bluff, and the place of these beds can be traced back under the soil-covered ridge to 857, but their stratigraphical relation to the rocks at 857 is very obscure.

On the S. W. ¼ Sec. 2 T. 34 N. 6 W. (861) occurs another outcrop, remarkable for the abundance and beauty of the fossil corals which it furnishes. The rocks embracing this deposit of corals I propose to designate as the "Acervularia Beds," and the particular stratum in which they are most abundant the " Coral Bed."

Section of Acervularia Beds (861.)

D. Shale, bluish, argillaceous, imperfectly seen at top of bank .... 2 ft.

C. Limestone, varying from dark to light gray, in beds from one to four feet thick, with a rough, somewhat granular fracture. Considerable argillaceous matter in the upper part. Few fossils.......... 23 ft.

B. Limestone, light or yellowish buff, varying to dark chocolate, argillo-calcareous, breaking with smooth fracture into irregular, sharply angular fragments, rather even-bedded in layers 6 inches to 2 feet thick. In the upper part alternating with bands of black bituminous calcareous shale and blue clay, 6 to 12 inches thick. The clay beds abound in beautifully preserved corals— *Acervularia Davidsoni*, *Favosites Alpenensis*, *Zaphrentis Traversensis*, &c. The bituminous bands burn quite freely, and have frequently passed for deposits of mineral coal .... 17 ft.

A. Limestone, grayish brown, compact, argillaceous,
uneven-bedded, with smooth conchoidal fracture, embracing in its upper part a 4 inch
stratum of black bituminous argillaceous limestone replete with *Atrypa reticularis* and considerable numbers of *Stromatopora* and *Acervularia Davidsoni* ....

The following additional species occur at this locality in Bed
B : *Favosites explanatus, F. mammillatus, Callopora, Fistu-
lipora* (2 sp.) *Aulopora* (2 sp.) *Trematopora, Fenestella, Spir-
orbis, Crania crenistria, Crania* (sp ?) *Aulopora cyclopora,
Cystiphyllum Americanum, Atrypa* sp? *Tropidoleptus* (3 sp.)
*Terebratula Romingeri, Strophodonta demissa, Cyrtia Ham-
iltonensis, Spirigera concentrica, Spirifer mucronatus, Spirifer*
(3 unrecognized species), &c., &c.

The stata embraced in the above section seem to be the equivalent of the eminently fossiliferous and often argillaceous
beds well known at other localities, as Partridge Point in
Thunder Bay, Widder and Saul's mills, in Canada West, and
Eighteen-mile creek in New York. Their stratigraphical cor-
relation to the other outcrops along the shore of Little Traverse
bay is less obvious, as the coa t line runs along nearly in the
strike of the formation. As, however, marked lithological and
paleontological distinctions exist at the different exposures, it
becomes necessary to decide upon their actual order of sequence, however provisionally we do it. The difficulty is en-
hanced by the presence of numerous undulations of the rocks
in the direction of their strike. At the locality under consideration the dip is toward the east or northeast, so that the strata
would be carried under 857; but there are sufficient grounds
for presuming that they rise again and pass *over* 857. Indeed,
the resemblance between the rocks at 857 and the lower part
of 861 is sufficiently exact to indicate this superposition.
Moreover, we have found on the opposite side of the State a
similar superposition, where the coral beds of Partride Point
hold a place above the bituminous Tropidoleptus Beds of
Thunder Bay Island.

So far, therefore, we may be justified in assuming the following order of superposition of strata of the Hamilton group.

. 3. Acervularia Beds, 858, 861.
2. Tropidoleptus Beds, 857. Bear creek. Saw-mill.
1. Stromatopora Beds, 856.

Rocks apparently identical with those of the last locality outcrop again at 862—S. W. ¼ Sec. 4 T. 34 N. 6 W. The coral clay bed is thinner and less exposed. The following is the section :

Section at 862.

C.	Limestone, buff, massive but shattered, crinoidal stems abundant in the upper part............	15 ft.
B.	⌈ Coral clay.............................:.....	8 in.
	⎨ Limestone as above the clay..................	2 ft.
	⎩ Lignite, calcareous and earthy, one to..........	3 in.

Limestone, buff, thick-bedded, shattered, with bands of lignite. Contains *Acervularia, Stromatopora* (wide cells), *Zaphrentis, Favosites, Cyrtia, Tropidoleptus,* and *Gomphoceras*.... 15 ft.

A.   Limestone, dark, fine, compact, thin-bedded, with conchodial fracture, extends under the water.. 12 ft.

The dip here is quite rapid toward the west, but the strata rise again in the distance of about half a mile, at 863, where we find a section as follows, embracing a dome-like elevation of the strata, from the summit of which they dip in opposite directions :

Section at 863.

⌈ Limestone, buffish, broken. Contains *Stromatopora* and *Acervularia,* and in the upper part, nu-
B. ⎨ merous crinoidal joints. (Thickness not noted).....................................
⎩ Limestone, similar to above, thick-bedded....... 4 ft.

A.   Limestone, dark, fine, compact, thin-bedded, breaking with conchoidal fracture. Extends under water.............................. 8 ft.

Thirty rods further west these strata arch up again, disappearing finally with a westerly dip.

The next outcrop—at 865, Sec. 1 T. 34 N. 7 W.—presents strata of very different physical characters, in a bluff about a mile long, and rising about 20 feet above the water level. The whole thickness of strata exposed is about 41 feet. The formation presents an undulating section, showing not less than four considerable synclinal axes, and finally disappears with a westward dip.

Section of the Buff Magnesian Beds (865.)

E. Limestone, brownish-buff, magnesian, arenaceous, moderately coherent, vesicular, thick-bedded, more grayish in the upper part, contains a few casts of shells in the lower part.............. 15 ft.

D. Limestone, darker colored, somewhat argillaceous, in broken layers from 1 to 4 inches thick. Contains *Nuculites*.............................. 6 ft.

C. Limestone, brownish-buff, magnesian, silico-argillaceous, porous, vesicular in streaks, in beds from 1 to 2 feet thick. Contains a band of *Favosites*. Reaches to water level at east part of the exposure ; further west is succeeded by the following :... 15 ft.

B. Shale, calcareous, soft, blue.................... 5 ft.

A. Limestone with *Acervularia* comes to view only at one point....................................

There is great difficulty in deciding upon the stratigraphical position of this section. I think, however, it lies above any rocks thus far described, because rocks which I assume to be the Acervularia Beds are seen to be beneath these thick-bedded strata at one point in the section. Moreover, the rocks in question finally disappear with an eastward dip, and range themselves apparently but a short distance below the uppermost strata exposed near Antrim.

The next exposure on the bay shore is seen at Pine River Point—880, N. W. ¼ Sec. 28 T. 35 N. 8 W. A limestone reef extends around the point, and just on the south of the point the rock rises in a broad swell, affording the following section :

Section at Pine River Point (880.)

D. Shaly bituminous bands, corresponding perhaps to the Lignite of 862..........................

C. Limestone, containing *Acervularia, Tropidoleptus, Favosites, Zaphrentis, Strophomena nacrea* and a little *Stromatopora*....................... 4 ft.

B. Limestone, very shaly and thinly laminated, containing *Fenestella, Stictopora, Tentaculites, Trematopora, Chœtetes, Chonetes scitula, Tropidoleptus* (3 species), *Cyrtia Hamiltonensis, Spirifera Marcyi* (but with extended wings four inches broad along the hinge) *S. Marcyi* (typical) *S. mucronata* (4 inches broad), *Strophomena* (with regular clean ribs and flat dorsal valve) *Spirigera concentrica, Terebratula?*.................. 10 ft.

A. Limestone, thick or thin-bedded, dark, highly calcareous, with green stains. Contains *Atrypa*
reticularis, Spirifer mucronatus, Strophomena
nacrea, &c.. 5 ft.

It would appear that the body of this exposure is in the same
horizon as that at 857—the " Acervularia Beds," coming in
above. It is obvious that the general dip of the strata at this
point is toward Pine river, since the great abundance of *Acer-*
vularia on the shore between here and Pine river proves that
the " Acervularia Beds" at the top of the bluff pass under the
harbor of Charlevoix. This conclusion is corroborated by the
fact that, in travelling north, after passing that portion of the
beach on the north side, in which *Acervularia* most abounds
among the fragments, we succeed to enormous quantities of
hard, fine and sharply angular fragments, whose position is not
far above the " Acervularia Beds " (865, D). Mingled with
these, moreover, are worn fragments of the brownish-buff magnesian beds of 865. It appears, therefore, that the rocks dip
from both directions beneath the harbor of Charlevoix, and
that Pine river finds its outlet along a partial synclinal axis,
produced by local undulations of the strata.

The " Acervularia Beds " rise to the surface again at 881—
on the line between secs. 29 and 32 T 34 N 8 W ; and still again
at 882, W ½ sec. 8, T 33 N 8 W.

The last out-crop of the limestone is at 884, N E ¼, sec. 34,
T 33 N 9 W, Emmet county, at which the following exhibition of strata is observed :

Section of " Chert Beds " (884).

E. Limestone, gray, in beds one to two feet thick, very
hard, with cavities containing sulphide of iron,
and calcareous spar .... 11 ft.
D. Limestone, gray, in laminæ a quarter of an inch
thick, with intervening sheets and concretions
of chert. Contains a few *Favosites*............. 9 ft.
C. Limestone, brown, in beds one foot and less in thickness.. 4 ft.
B. Limestone, bluish, shaly .... 1 ft.
A. Limestone, bituminous, irregular, broken......... 1 ft.

The dip here is toward the southeast. The rocks seem to be
quite distinct lithologically, from any others along the coast.

The thinly laminated or shaly limestone beds (D) graduating into, or alternating with, thicker limestones, resemble the rocks at 880, B, but, if identical, the abundant fossils of the last-named locality would scarcely be entirely wanting. A very peculiar character is also imparted by the abundance of chert. The hard massive limestone above these beds is not seen elsewhere, and constitutes, so far as observed, the uppermost member of the Hamilton limestones, since the dip at this place, would bring it immediately beneath the "Black Shale," which makes its first appearance about a mile further south.

Putting together the observations made at the various points, we are enabled to arrive, for the first time, at some definite idea of the order of sequence of the various members of the group of Hamilton limestones. This sequence may be thus exhibited:

V.   Chert Beds, (884) ....24 ft.
IV. Magnesian Beds, (865)..........................35 ft.
III. Acervularia Beds, (858, 861, 862, 863) ....23 ft.
II. Tropidoleptus Beds, (857, lower part of 862, 880,
        881)..15 ft.
I.   Stromatopora Beds, (855, 856) ....44 ft.
                                                    ———
    Total....................................141 ft.

The Acervularia Beds probably find their equivalent in the upper part of the bluff at Partridge Point, in Thunder Bay, on the opposite side of the State, and again at Widder and Saul's Mill, in Canada West. The Tropidoleptus beds are found reproduced in the bituminous limestone of Thunder Bay island, and the lower falls of Thunder Bay river. The Stromatopora Beds are seen emerging from below the water level on the northeastern side of Thunder Bay island. The Magnesian and Chert Beds have not as yet been elsewhere identified—though we find a series of limestones near the upper falls of the Thunder Bay river, which have not yet been correlated with the other members of the group.

The Hamilton limestones sweep across the mouth of Grand Traverse bay and determine the general trend of the coast from Light House Point to near Frankfort on the Lake. The water at the mouth of the bay shoals to 10 or 15 fathoms, gradually increasing in depth southward in the direction of the dip of the rocks to 103 fathoms opposite Old Mission; while on the north, the increase of depth is more abrupt, corresponding to the sublacustrine escarpment. The limestones of this group form a reef about Light House and Cathead points, and again in Canfield's harbor. They form the basis of that tongue of land known as Carrying Point and of Bellow's island off the mouth of Northport harbor. It is probable that the same formation constitutes the foundation of the Manitou islands, since the depth of water between the North Manitou and Pyramid Point (or North Unity) does not exceed 25 fathoms in the deepest part, and shoals abruptly in both directions to 7 fathoms and less.

5. SHALES OF THE HURON GROUP.—In following the shore from Little to Grand Traverse bay, we find the "Chert Beds" succeeded, within the distance of a mile, by an outcrop of bituminous shale. This first appears at 888—Sec. 3 T. 32 N. 9 W., Antrim county. It rises toward the south and finally attains a thickness of about six feet. It is a rather hard, black bituminous shale containing iron pyrites, and, in the lower part of the exposure, some bands of silicious shale looking like a recurrence of the thin chert beds of 884. On exposure, the shale cracks into multitudes of small angular fragments and finally disintegrates.

This bed can be traced in a bluff a little retired from the shore and covered with soil, to 889, about forty rods distant, where it again outcrops in a bluff about 15 feet high, presenting nearly the same lithological characters as before. The oxydation of the pyrites, on exposure to the air, produces a reaction which forms a whitish efflorescence of sulphate of iron, or copperas, on some of the exposed surfaces, and thus greatly disguises the real characters of the rock.

The same formation outcrops again at 890, which is only about 20 rods south of 889. The lithological features remain the same, though the entire thickness of the black shale here exposed is about 20 feet—some enthusiastic, but deluded, searchers for coal having opened an excavation about six feet in depth.

4

At 891 — S. W. ¼ Sec. 11 T. 32 N. 9 W.—is another and important exposure of this formation, giving a view of about 20 feet in thickness. The shale for an interval of about 40 rods exhibits evidences of some geological disturbance. The strata are sometimes abruptly broken into huge angular blocks standing at all angles. It is difficult to decide what agency frost may have had in producing these dislocations. At a spot a few rods further south, however, are seen two narrow, nearly vertical fissures, now filled with calcareous spar. The whole mass of shale is intersected by divisional planes, making an angle of 70° with the stratification. No dip appears along the general face of the section, but at a notch in the coast line there is seen a dip of 6½° toward N 51° E, which is nearly in the direction of the strike of the formation. It will be observed here as elsewhere throughout the peninsula, that the normal dip of the formation is quite imperceptible, and the only dip which can be detected is merely local, produced by the undulations of the rocky sheets, and may be in any direction whatever.

From this shore the bituminous shales strike toward the northeast, and are next known on the north side of Pine lake, about 6 miles from Pine river dock. The locality—868— N. E. ¼ Sec. 3 T. 33 N 7 W.—is about 20 rods back from the shore of lake Michigan, at a point about one mile south of Frankfort, in Benzie county, as I have been informed. In the intermediate distance no actual outcrop is at present known. As these shales, however, are extremely friable, and their fragments could not bear the violence of a prolonged transportation, I consider it sufficiently exact for a provisional determination, if we assume the place of outcrop of the formation to lie along the belt of most abundant surface fragments. Relying upon this criterion, we are led to infer that the black shale strikes the west shore of Grand Traverse bay, in the neighborhood of New Mission, and passes thence southwest to Carp lake, which it crosses about a mile below " the narrows." It is next reported about six miles east of Glen Arbor, but I cannot vouch for the statement. I have, however, seen fragments to the southeast of Glen lake in the presumed trend of the formation.

The black shale thus traced is the same formation as that in Thunder bay on the opposite side of the State, at Kettle Point, and at various other localities in Canada West. It is the equivalent of the " black slate " of Ohio, Indiana, Kentucky and Tennessee—extending even into Alabama. It is the Genesee shale of the New York geologists. Its identity with the Genesee shale rather than the older Marcellus shale with which it was formerly identified, is established by its stratigraphical position both in Grand Traverse and Thunder bay. It is ranged immediately above the limestones of the Hamilton group instead of below them, in the place of the Marcellus shale. Moreover, the black shale of the Huron group, though almost uniformly destitute of marine fossils, has at length afforded me a few specimens from near the mouth of Bear Creek, in Canada West. Among these I identify *Discina Lodensis* and *Leiorhynchus multicosta*— species known to be restricted to rocks above the Marcellus shale in the State of New York.

The black shale of the Huron group is known to be succeeded in ascending order by a great thickness of whitish or greenish and more or less calcareous shales and clays ranged under the same group, though thus far totally unproductive of fossils for the determination of their affinities. The most conspicuous outcrop is seen on the east shore of Grand Traverse bay at 893 —Sec. 36 T. 32 N. 9 W., Antrim county — extending thence southward for half a mile or more. It seems to be formed by a gentle swell of the formation, with minor subordinate undulations. The rocks are a calcareo-aluminous shale, occurring in layers from half an inch to two or three inches in thickness. In some portions of the exposure, the layers are somewhat arenaceous, and at times assume the characters of a shally argillaceous sandstone. The whole thickness exposed is about 15 feet. Two noticeable folds occur at this exposure within 30 feet of each other, and a third a few rods further north. The first and last present each a downthrow of about a foot. The middle one is much the greatest, presenting a downthrow of about four feet. These dislocations are not properly faults, for the strata are not fissured, but rather folded as if by a powerful lateral pressure. This is as great a disturbance of the strata as has been noted in the lower peninsula—a similar one occurring in rocks of nearly the same age in the neighborhood of Pt. aux Barques.

The green shales strike southwest across the bay, but no actual outcrop of rocks of this character has been observed in that direction. A reef of light calcareous shale exists off Mission Point, and, judging from knowledge of the formation obtained in other parts of the State, the position of this reef is probably above the green shale, but not far removed. Striking diagonally across the " Peninsula," the light calcareous shale appears just beneath the water level on the north side of Tucker's Point, —Sec. 17 T. 29 N. 10 W.

I have not discovered the means of tracing the green and light colored shales any further. It may be assumed, however, that their strike continues southwesterly along a line nearly parallel with the outcrop of the black shale. Their thickness is not adequately indicated by the few exposures accessible, since, in other portions of the State we find it to reach four or five hundred feet.

It is probable that this series of greenish and light colored, argillaceous and arenaceous shales corresponds to the Portage group of New York. If the overlying Marshall sandstone should finally be shown to occupy a position above the Chemung group of New York, it will become necessary to admit that the shales under consideration embrace both the Portage and Chemung groups of New York. In this case, the Huron group, as orginally defined, will extend from the bottom of the Genesee shale to the top of the Chemung group.

6. MARSHALL SANDSONE.—No stratified rocks higher in the series than the light shales have been observed within the limits of the Grand Traverse region. There are, however, geological reasons for believing that the southeastern portion of this region is underlaid by the buff, and friable sandstone of the Marshall group, which immediately succeeds the Huron group. Nothing more than an approximate indication of the boundary of this sandstone can be made; and this has been attempted on the map, by drawing a line so as to cut off nearly

the whole of Kalkasca county, and the southeastern angles of Antrim and Grand Traverse counties.

7. Drift Materials.—All parts of the Grand Traverse region, like other portions of the lower peninsula, are buried beneath accumulations of sand, gravel and clay, entirely destitute of a stratified arrangement, or presenting only a confused or irregular stratification. These deposits are the product of geological agencies that have been at work during the last period of the world's history. Their average thickness in this region is unknown. It is probable, however, that they are 50 feet thick at Northport, 60 at Sutton's bay, 100 at Traverse City, and from two to four hundred in the interior of Leelanaw and Benzie counties.

If we examine the structure of these deposits, we find the surface generally composed of sand, with occasional belts and patches of clay. The sandy constitution extends downward a varying depth, sometimes 50 or 100 feet; but we always encounter, sooner or later, one or more beds of clay. The clay deposits are in the form of vast sheets or basins, inclining at all angles, overlapping each other in various ways, and disposed at various depths, with sand both above and below. The bottom of the drift accumulations, however, is made up, generally, in this, as in other regions, of an enormous bed of clay, pebbles and boulders, resting on the outcropping edges of the rocks.

If we examine the mineral constitution of the Drift deposits, we find that most of the boulders and pebbles of the underclay are derived from granitic, syenitic, dioritic, quartzose and gneissoid rocks, and micaceous, talcose, hornblendic and silicious schists. No such rocks are found in place within 150 miles. These fragments have been transported from the upper peninsula of the State. We find corroboration of this opinion in the discovery of masses of native copper mingled with the other materials. One such mass found near Northport weighed —— pounds, and was sold for eighty dollars. If we examine the fragments of rock disseminated through the arenaceous and more superficial portions of the Drift, we discover, from their mineral character and their fossil remains, that they have been to a great extent derived from the rocks immediately underlying, or outcropping but limited distances towards the north. The coarser Drift materials are, therefore, partly of local and partly of foreign origin. The source of the fine sand and the fine argillaceous deposits is somewhat more obscure. Without attempting to elaborate the evidences, it is enough to state that the boulder clay is believed to have had a northern origin, while the fine sand may have been derived from rocks of various ages, removed to various distances from their place of deposit, but is believed to have been mainly derived from arenaceous limestones of the Hamilton, Corniferous and Onondaga salt groups. The innumerable fragments of these limestones — especially the Hamilton — disseminated through the soil and subsoil, have been for ages undergoing a slow decomposition. The calcareous matter escapes in a state of solution and affords an important fertilizing constituent of the soil, while the imperishable grains of sand loosened from their bonds by the solution of the calcareous cement, become a principal portion of the finer material of the soil.

It is a very general opinion that the ruggedness of some portions of the Grand Traverse region is caused by disturbances of the underlying strata. I have, however, failed to discover any correspondence between the configuration of the surface and that of the underlying rocks. The hills are mere piles of Drift materials. The Drift was originally left with an uneven surface, but the depressions have been subsequently further scooped out by the erosive action of rains and torrents. The same agencies are continually wearing down the hills by removing the finer and looser materials to the valleys. If the configuration of the hills be attributed to uplifts of the underlying rocks, this is to suppose that the underlying rocks have at some time undergone a great degree of disturbance — much greater than the appearance of the rocks would warrant at any place where we have been enabled to inspect them. We know that everywhere else the strata of the lower peninsula repose in nearly horizontal planes. It is only in eruptive regions like that of Lake Superior, that we find the rocks forming the backbone of the hills.

But we need not speculate on the constitution of the hills. The erosion of lake Michigan along the eastern shore has gnawed away the land, till in some instances, the water-line has been carried to the very heart of the highest eminences in the country. The sloping lakeward faces of Mount Carp, North Unity, Sleeping Bear Point, Empire Bluff and Pt. aux Becs Scies are natural sections through the highest parts of some of the hills. In every case they fail to disclose any rocky ledges, but on the contrary, exhibit accumulations of pebbles, sand and clay to the very water's edge.

The topography and vegetation of Sleeping Bear Point have been already described; and I have just stated that this prominent land-mark from the lake is merely an enormous pile of diluvial rubbish. Toward the base, thick bands of pale bluish and purplish clay crop out, separated by beds of gravel and sand. Some portions of the surface also expose masses of gravel rendered adhesive by an intermixture of clay. Here and there a huge boulder protrudes above the general suface, polished like the smaller fragments, by the incessant pelting of sand particles driven before the wind. Much of the plateau is strewn with small angular fragments of chert; and it was long before I accounted for the preservation of their sharp angles among deposits that have suffered so much from attrition. I discovered at length a few large boulders of chert-bearing limestone—apparently from the Chert Beds of the Hamilton group, as I subsequently learned in which imbedded masses of chert had been shattered *in situ*, perhaps by the action of frost. The dissolution of the inclosing limestones loosens the cherty chips, and the winds and rains strew them over the bald surface of the plateau.

The beds of clay at the bottom of the Drift are frequently found so free from pebbles and so evenly stratified as to be easily mistaken for some member of the argillaceous series of the Huron group. This is the case with some of the beds outcropping in the high beach of lake Michigan, between Leland and Cathead Point, as also at North Unity and Sleeping Bear, and on the west side of the Peninsula south of Bower's harbor. In other cases the similitude of an older formation outcropping *in situ* is sometimes assumed by beds of fine sand either cemented by carbonate of lime, so abundant in the soil, or simply stuck together by clay. Phenomena of this kind are observed in the bed and bank of a small creek at Provemont, and also on the east shore of Carp lake, north of " the narrows." The latter kind of formation is comparatively recent, as is proved by the presence of inclosed stems or leaves of modern vegetation, or the shells of fresh water molluscs, or by the occurrence of uncemented Drift beneath them. The former kind of formation may be assigned to its true position by observing whether, in any part of its extent, it embraces water-worn pebbles, or presents great and abrupt variations in constitution, in the regularity of its stratification, the thickness of the separate layers or the persistence of the dip.

8. LIGNITIC DEPOSITS.—At numerous places in the Grand Traverse region we find accumulations of vegetable matter and silt presenting a brown or blackish color, and occuring under a somewhat stratified arrangement. These accumulations occupy a position above all the stratified rocks, and the indications are that they are of more recent date than the boulder clays. At the same time we often discover thick deposits of sand, clay and shingle resting above them. They occur at various elevations, from a depth of eight or ten feet beneath the water level of lake Michigan to the height of fifty feet above. The most noteworthy instances will be cited.

In the neighborhood of Brownstown, at the southern termination of the green shale already described, occurs the following series of strata:

Section of Lignitic Deposit.

F.	Fine yellow sand—the subsoil of the region.......	12 ft.
E.	Small boulders, pebbles and coarse sand with shells of *Melania* and *Physa*......................	7 ft.
D.	Clay, soft, arenaceous and bituminous............	2 ft.
C.	*Lignite*, somewhat impure, containing stems of cedar and other exogenous vegetation, passing above and below into a more argillaceous state..	3 ft.
B.	Clay, dark gray, very tough, with some sand and small pebbles and bituminous matter..........	2½ft.
A.	Green clay, appearing to be produced by the disintegration of the green shale which holds a position immediately below, though not in juxtaposition,	2 ft.

The clay, lignite and green shale lie nearly in the same level, and their succession is made out only by carrying the observations laterally for a short distance. The lignite beds are spoken of as bituminous. They are so to some extent, but most of the vegetable matter is rather in a peaty or carbonaceous condition.

The foregoing observations were made in 1860. At the present time a large part of the exposure is covered up by sand, which has run down from above. But, on the other hand, we were now enabled to make observations which were not made five years ago. The lignitic bed is found passing under the sand which intervenes between the bluff and the water's edge, and can be seen beneath the water at a distance of three rods from the shore. Moreover, I was informed by a fisherman, that they penetrate it in driving their stakes for pound nets where I saw them, at the distance of a third of a mile from the shore, in water said to be eight or ten feet deep.

The occurrence of this deposit at such a depth beneath the water level, and at such a distance from the shore, renders it necessary to adopt with great caution, any explanation identifying it with an ancient accumulation of drift wood stranded on the beach. At the same time there are not sufficient evidences of its sedimentary origin; especially since, at other localities, a similar formation is found at considerable elevations above the lake.

On the east side of Carrying Point, near Northport, I observed a similar deposit at the water's edge, and extending a few inches beneath the surface. This rests on the undulating surface of a shingle beach, and, in one or two places, is seen to extend back into the borders of the forest, passing under the recent accumulations of leaves and shrubbery, and presenting the ordinary characters of a peaty deposit.

Again, on the sloping face of Sleeping Bear Point, in the vicinity of the "Little Bear," are seen bands of dark, lignitic matter, forming irregularly disposed belts along the exposed section. On examination, some of these are found to consist of sand, mingled with peaty particles. Above and below is blown sand, and the whole mass is apparently a mere dune formation—the peaty particles assorted out by eddies of wind. These peaty particles, however, had their origin in the turfy soils with which some portions of this Point have, at some time, been covered, and relics of which are found still preserved, and presenting, in other places along this bluff, outcropping beds of better characterized lignitic material.

Similar accumulations of peaty or lignitic matter are exposed by the erosion of most of the streams emptying into Carp lake from the west; also about a mile northwest from Traverse City, in the bed and banks of a creek; also on Peter Stewart's land, Sec. 17 T. 29 N. 10 W., on the Peninsula; also at Whitewater, Sec. 21 T. 28 N. 10 W., on the land of A. T. Allen. At Northport, in the bed of the creek back of the Traverse Bay Hotel, is a mass of bedded peat containing fresh water shells, and, in places, becoming marly. In another situation, and deprived of its shells, this deposit would pass for lignite; but in this situation, with springs oozing out of the banks, and fresh water shells so abundant, the deposit can scarcely be regarded as anything different from ordinary peat.

From the observations made on the lignitic accumulations of the Grand Traverse region, I am led to think that the principal deposits are not sedimentary accumulations formed in the bottom of the lake (and bay) near the shore; nor masses of stranded drift vegetation; nor materials bedded in the Modified Drift by either marine or lacustrine action at heights above or below the present water line; nor do I think changes of level in the lake waters have had any connection with their occurrence above or below the present water level. They seem to be ancient peat beds formed in situations kept moist, in some cases, by access of water from the lake, in others by the percolation of spring water from contiguous sand banks. They may hence occur at any elevation above the water level, and present exact adaptations to the inequalities of the subjacent surface. The erosion of the lake and bay shore has caused these peaty areas to be invaded by the waves, which, while they could not, without unusual violence, rend to pieces the peaty matter bound together as it is by interlacing stems

and fibres, could nevertheless wash out the fine sand on which
the peat bed rested, and cause it, by degrees, to settle down
to the water level, and even beneath it.

XI. ECONOMICAL IMPORTANCE OF THE VARIOUS GEOLOGICAL FORMATIONS.

1. SALT.—The Onondaga salt group, which underlies the
entire region, is the source of supply of the gypsum and brine
of central New York. It furnishes the gypsum of the Grand
River region of Canada West, and of Sandusky bay in Ohio.
It is also known to contain a large supply of gypsum in this
State at Little Pt. au Chene, west of Mackinac. I have also
shown that the salt wells of Port Austin and St. Clair, in this
State, are supplied from this source; and have expressed the
opinion that this group of rocks will be found equally productive in other portions of the lower peninsula. The position of
the Grand Traverse region is such that I should be led to hope
for success in boring into this formation. The well authenti-
cated existence of an ancient salt spring on the neck of land
connecting Harbor (or Hog) island with the Peninsula, I should
regard as a confirmation of this opinion, since, if a fissure ex-
isted in the overlying rocks, the brine would tend to rise by
hydrostatic pressure, as through an artesian boring. Deacon
Dame, of Northport, one of the oldest residents of the region,
has furnished me with detailed information, which seems to
fully authenticate the current tradition relative to the former
existence of this spring.

Mr. H. G. Rothwell, of Detroit, likewise informs me that a
salt spring exists on the southwest corner of Sec. 35 T. 26 N.
16 W., which is less than three miles from Frankfort, in Benzie
county. Undoubtedly this spring is supplied from the same
source.

Very great difficulty exists in estimating the depth from the
surface at which the formation would be struck. If an experiment were to be made at the head of the East or West bay—
points where the basin would be found most depressed, and
the brine, consequently, most concentrated—we might venture

to make the following approximate estimate of the thickness of the intervening formations:

Drift Materials	120 feet
Light and Black Shales	400 "
Hamilton group	140 "
Corniferous limestone	200 "
Onondaga Salt group	50 "
Total	910 "

The light and black shales (Huron group) attain a thickness, in the southern part of the State, of about 600 feet, but I believe the indications do not justify so high an estimate for the Grand Traverse region. The Corniferous limestone is about 300 feet thick at Mackinac, but not over 100 feet thick in Monroe county. I think the question of salt would be sufficiently tested within 1000 feet.

2. PETROLEUM.—The Hamilton group is the formation in which most of the oil is obtained at Oil Springs, Petrolea and Bothwell in Canada West. It consists there, as here, of a series of limestones, shales and shaly limestones. The oil accumulates in the loosely constituted shaly limestones; in the numerous small fissures of fissile clay shales; in vertical fissures and irregular cavities in the massive limestones, and in the pores of a buffish, porous magnesian limestone at the bottom of the series. The deposits of oil possess no considerable lateral extent, since wells even upon the same acre of ground seldom interfere with each other. The oil also accumulates, sometimes in enormous quantity, in a bed of gravel or sand reposing at the bottom of the Drift materials upon the top of the rock. This is a thick oil used for lubricating purposes.

At Petrolea the black (or Genesee) shale is not found overlying the rocks of the Hamilton group. At Oil Springs the thinning edge of this formation is encountered in about the middle of the productive area. This thickens toward the west, until, in some wells not over a mile distant, the formation has attained a thickness of 40 to 50 feet. At Bothwell the black shale occurs of considerable thickness. The same is also true of the undeveloped regions of Wyoming, Dawn and Chatham in the peninsula of Canada West.

There are belts within the Grand Traverse region corresponding in geological position with each of the localities just named; and I believe there are good geological reasons for anticipating success in an attempt to obtain oil. The region from the head of Little Traverse bay to Northport, and thence to Leland, Glen Arbor and Frankfort, is situated like the region about Petrolea, except that the drift materials, inland from the lake shore, are accumulated in deeper masses. The best situation for making experiments would be at points sufficiently covered to have prevented the evaporation of the oil, but yet sufficiently depressed to avoid unnecessary boring through the overlying sands.

The line which marks the western boundary of the Genesee shale—already indicated—marks out a belt of positions similar to that of Oil Springs in Canada West; while a strip of country a little further east would be found circumstanced similarly to the Bothwell oil region.

It cannot be expected that oil will be found generally and indiscriminately distributed throughout this area, but I should be surprised if half a dozen undertakings, judiciously located, should fail entirely of bringing the coveted fluid to light.

Surface indications are quite common throughout the region, of which the following are a few examples.

In a stream at Lindsley's house, at Sutton's Bay. Mr. de Belloy and I have demonstrated that this proceeds from a marsh half a mile distant on the hill-side. Strong indications exist also on the farm next north of Lindsley's.

On Carp lake, near the landing of Cornelius Jones, I saw a film of oil, and brought up bubbles of inflammable gas by stirring the bottom.

On the east side of the lake, a little further down, I saw similar indications. I saw the same again near the head of the north arm of the lake. Again, on the back part of Buckman's farm, and throughout that region, north of Leland, on low grounds.

On the west side of Carp lake, near the head, on land of James Nolan, I noticed indications; and also at numerous points between Nolan's and Provemont. No smell was noticeable, and in many cases the oil was mixed with an iron film.

Mr. McPherson, living on the east side of the bay, about four miles south of Antrim, assured me that he had detected a strong smell of kerosene in passing the outcrop of green shale in that vicinity. One of his boys asserted that he found the smell so strong, one day, that he hunted long for the fragments of the jug of kerosene which he was convinced had been broken at the spot. Mr. de Belloy gives similar testimony to the occurrence of a strong smell in the same vicinity, at certain times. The same is also asserted by Mr. Blakely, living near Torch lake.

About two miles from Northport, on the road to New Mission, I saw a fine film of oil on standing water.

In the border of the swamp back of Deacon Dame's residence at Northport, are very characteristic indications. Gas also escapes at intervals in a spring near the house. In calm weather a copious escape of oil can be seen from Rose's dock. The oil rises and spreads in a fine film with dichroic refractions on the surface of the water. The same is seen again at the mouth of the creek near the dock.

Similar phenomena are seen on Manseau's creek at Pishawbey-town. So says Mrs. Page.

On the land of Rev. Mr. Smith at Northport, I saw abundant films of oil, with some iron. The same can be seen on standing water near the creek back of Traverse Bay Hotel.

On the northeast shore of Leg lake, I saw several small oil springs, with much iron.

On land of Rev. George Thompson at Leland, I saw slight petroleum indications near the house, and also about a cattle spring a quarter of a mile northwest of the house.

Supposed oil indications occur on railroad section 3 T. 32 N. R. 7 W., about a quarter of a mile from the north side of Pine lake.

In the low ground about the head of the West Arm of the bay, I noticed abundant films of oil on the surface of standing water.

The numerous instances in which the escape of oil and gas to the surface has been observed, tend to confirm very strongly the induction based on stratigraphical data, and afford full justification of attempts to reach, by boring, the reservoirs whence the oil escapes.

3. CLAYS.—The Drift formation, besides supplying an admirable quality of silicious sand for plastering, contains large deposites of pure clay for bricks and pottery. In some cases this clay is already mixed with a sufficient amount of sand for immediate use. Along the elevated beach north of Carp river, is an exposure of an enormous deposit of fine fawn-colored clay, quite free from pebbles of every sort. It is compact and somewhat fissile, but undoubtedly belongs to the drift formation. The same deposit outcrops again at North Unity, at Sleeping Bear and at Empire Bluff. At either point a manufacture of bricks could be established which would rival Milwaukie both in the cheapness of production and the fine quality of the bricks. Beds of excellent clay occur at Frankfort within the limits of the town plat.

Clay of similar quality, but somewhat mixed with boulders, occurs on the bay shore south of Antrim. A bed of boulder clay abuts upon the bay at New Mission, and forms the basis of the promontory on which the Seminary stands. A short distance back of Fisher's house at Glen Arbor, a land slide has uncovered an excellent bed of pure clay. At Antrim a second beach, a few rods inland, is formed by a bank of pink-colored, boulder-bearing clay.

It is a mistake to suppose the clay of the region is not adapted to brickmaking. No doubt limestone pebbles may become mixed with the clay employed, but a good article can be successfully selected. William Wilson informed me that he made 2,000 bricks from a bed of clay two miles below New Mission, and they proved unexceptionable. The color was that of Milwaukie bricks. Of the whole quantity not one has bursted from the presence of limestone pebbles. At Elk Rapids 100,000 bricks of Milwaukie color were made, and all were good, except a few made from material taken near the surface of the bed.

XII. FARM CROPS.

The descriptions which I have given of the nature of the soil and climate of the Grand Traverse region, will prepare the reader for the statement which I now make, that the region is capable of producing any crop which flourishes in the northwestern States, as far south as the latitude of Cincinnati. In order to give definiteness to the testimony which I am about to produce, I shall furnish names, localities and figures. Where anything is given on the authority of others, the production of their names will render them responsible for the statements.

1. WHEAT.—The staple crop of the region, at present, is winter wheat. The mildness of the autumn enables it to secure a good start, while the mantle of snow with which the country is covered during the winter, insures the crop against winter-killing. Very rarely I heard accounts of " smothering " in limited localities.

Mr. Hannah informs me that the average production of wheat about Traverse City is 25 to 30 bushels per acre. Morgan Bates, Esq., says the first crop of wheat pays for clearing the land. In 1864, he cleared 27 acres of heavily-timbered land, hiring all the labor. The clearing, fencing with temporary fence, seed, plowing part, sowing, harvesting and threshing, cost $892. The wheat raised was 560 bushels, which sold at Traverse City at $1.60 a bushel, amounting to $896. This yield is only 21½ bushels per acre, but Mr. Bates states that an unusual amount was wasted by improper harvesting. The land is now worth $30 per acre.

Rev. Merritt Bates published the following statement in 1863: Cost of clearing ten acres, fencing, seed, cultivation, $285. Product, exclusive of waste caused by threshing in the open air, 268 bushels, worth at the door $1.25 per bushel, $335 —besides straw worth $50.

James Orr, one mile south of Antrim, raised winter wheat, a sample of which was stated by dealers at Battle Creek to be the finest known in the State. William Johnson, on the east side of Elk Lake, raised this year 30 bushels of winter wheat to the acre. In the Monroe settlement, in Grand Traverse county, winter wheat averages 28 bushels to the acre. In 1863, William Monroe raised 30 bushels, and his brother Henry 38 bushels to the acre.

2. CORN.—The variety of corn most prevalent is the yellow eight-rowed corn. It has been demonstrated, however, that dent and King Philip corn and other varieties, will ripen with certainty. I saw Mr. de Belloy and others husking corn fully ripe in the middle and latter part of September, and I was assured that the crop was sufficiently ripe quite early in September. I saw fields of corn fully ripe about the same time, near the north end of Carp lake, and also on the west side, toward the head of the lake. It was growing thriftily on the steepest hill-sides, in fields which, in some cases, had not been plowed. I saw dent corn fully matured and twelve feet high on the land of Stephen Perkins, near Long lake. At Leland, I saw ears of King Philip corn, raised by H. S. Buckman, which were 10½ and 11 inches long, well filled out and matured. I saw similar ears at Traverse City,—also luxuriant specimens of Ohio dent corn. Enoch Wood, four miles south of Elk Rapids, brought to market two loads of dent corn equal to any produced further south. Wm. Monroe informed me that he raises good crops of corn—some of which is dent corn. Dent corn was raised this year by E. P. Ladd, at Old Mission. Corn is not so sure a crop at Pine river.

3. OATS.—On the west side of Carp lake, I saw as good crops of oats as ever in my life. Mr. Hannah states that oats always bring an excellent crop. I saw in the office of Mr. Bates, at Traverse City, a bunch of oats 7 feet and 9 inches high, raised on land of John Cornell, 20 miles south on the Newaygo state road. The whole field is said to have been extra-ordinary, though I was assured that many farmers have raised fields of oats six feet high. Mr. Monroe informed me that he raises 50 bushels of oats to the acre.

4. BUCKWHEAT.—As might be expected, buckwheat also flourishes luxuriantly. I never saw better fields than on the west side of Carp lake. The crop is generally said to flourish well, but is not perhaps extensively introduced.

5. POTATOES.—The finest potatoes of the country are produced in this region. The soil and climate seem to be admirably adapted to the crop. Thousands of barrels are shipped to Chicago annually. They often pass in the market for Mackinac potatoes, as that region had a reputation established many years ago. The potato grows without cultivation in the Grand Traverse region—the entire crop being often left in the ground till spring, and scattered tubers taking root in fields cultivated for other crops. Mr. Fisher, of Glen Arbor, had potatoes growing in a field where they were planted six years ago.

The potato grows large and smooth and is uniformly healthy. I saw at Traverse City, a Carter potato, raised four miles west of that place, on land of Rev. Merritt Bates, and measuring 8¼ inches in length and 9 inches in circumference. In Campbell's store, at Northport, I saw three Peach-blow potatoes, weighing respectively 19½, 20 and 26 ounces. Deacon Dame informed me that he had raised 300 bushels of potatoes to the acre. He also says that a single hill sometimes yields considerably over a peck, and that whole fields will average a bushel to every eleven hills. This was done on land of Mrs. Daniel Knox, two miles west of Northport. Mrs. Page, of Pishawbey-town, says she raised from one hill a half bushel even full of Lady-finger potatoes. It was only one hill of a patch. Mr. J. W. Washburne says he raised on one stalk of Peach-blow potatoes, over a half peck—all large ones. This was in a half acre lot, the soil of which had been cultivated several years. W. W. McClellan, of Northport, showed me a potato of Clinton variety, raised on land of James Martin, 2½ miles north of there, which measured nine and seven-eighths inches in length and weighed 33½ ounces. Mr. Tilley, at Leland, showed me two potatoes of the Cazenovia variety weighing 18½ and 20½ ounces respectively, ten days after being placed in a dry atmosphere. They were raised two miles south of Leland. Mr. Gerard Verfurth exhibited a potato of the same variety, raised in the village, which weighed 27 ounces. I saw potatoes of the California variety, raised by Rev. Mr. Smith, of Northport, measuring 8½ inches in length, and a black Meshannock 8¾ inches long.

6. HAY.—Timothy hay proves a successful crop. Mr. Bates, of Traverse City, has 33 acres seeded, which he calculates will pay him the interest on $3,000. If it brings only one ton to the acre, he will receive a profit of $9 per ton, or $307 on the whole, which is ten per cent. on $3,070, or about $93 per acre. Rial Johnson, four miles south of Elk Rapids, has one of the oldest farms in the country, and raises superior Timothy hay. Mr. E. Pulcifer, south of Elk Rapids, got 19 loads of red clover hay from three acres planted to an orchard. He keeps nine cows, and makes butter and cheese for the market. He proposes to enlarge his dairy. I saw first rate Timothy hay in the fine, capacious barns of William Monroe, in Grand Traverse county.

7. OTHER CROPS.—Turnips grow with the utmost luxuriance, as I have observed on the west side of Carp lake, at various places about Traverse City, and along the road thence to Glen Arbor. Mr. Sprague, near Leg lake, in Leelanaw county, directed my attention to a fine field of turnips, and assured me that he once raised a flat turnip which weighed 17 pounds (!) Carrots grow well. Mr. Stewart, on the Peninsula, showed me a bed of carrots which were from two to three inches in diameter, the seed of which was planted July 1st. He showed me parsnips of a still larger size. I saw fine carrots back of Glen Arbor. Tomatoes ripen well. I met with them at various points. Mrs. Joseph Batey raised three tomatoes in the south part of Traverse township, which weighed 40 ounces each. Mrs. Dixon informed me that tomatoes do not mature well at Pine river. I saw an excellent crop of white beans at Rial Johnson's; and also large, plump marrowfat peas.

XIII. FRUITS.

As a fruit-growing region, it is doubtful whether any other part of the United States will compete with this. Apple trees were planted on the first settlement of the county, and have always grown well and borne luxuriantly. The characteristics of the trees and fruit are healthfulness, luxuriance and large size. Rev. Mr. Smith, of Northport, has a young orchard in which I saw various familiar varieties in a greater degree of perfection than in any other part of the country. The average size of the Rhode Island Greenings was eleven inches in circumference—weighing eleven ounces. Seedling apple trees were loaded with fine winter fruit. Fine young orchards are coming into bearing on all parts of the Peninsula, and throughout the country south and southeast of Elk Rapids. Mr. Hannah, at Traverse City, has planted an orchard of about forty acres containing 1,000 trees. At New Mission, I witnessed the most beautiful exhibition of apples that ever met my eyes. An orchard on the seminary grounds, about 14 years old, was completely loaded with large, fair, richly-colored fruit of old and new varieties. It was a marvel of luxuriance and beauty. I saw whole trees borne down with apples from four to four and a half inches in diameter, and weighing from 14 to 18 ounces. These trees were planted and reared by Rev. Peter Dougherty, the intelligent and useful superintendent of the mission. I saw young apple trees flourishing luxuriantly in the neighborhood of Glen Arbor, and in nearly all other parts of the region.

It was formerly supposed that the climate was unsuited to peaches, but different persons having from time to time planted a few peach stones, it was ultimately proven that the peach flourishes in perfection. At Leland, I saw trees laden with ripe fruit in September. At New Mission, the peaches which I saw were as great a marvel as the apples. Some measured eight and nine inches in circumference. The seedling fruit was so abundant that no attempt was made to gather it. Thomas Tyre, on the Peninsula, brought to market this year 75 to 100 bushels of peaches. Rial Johnson, on Elk lake, raised 200 bushels from a small orchard, the seeds of which were planted ten years ago. Rev. Mr. Smith, of Northport, succeeds with peaches. Mr. Fisher treated me with peaches raised at Glen Arbor. I saw thrifty trees growing on the farms back of Glen lake. Mr. Almon Young on the south side of Round lake, raised superb peaches; also Mr. Amos Wood, two miles from Elk Rapids. Mr. Wood's trees have been bearing regularly for six or seven years. Mr. E. Pulcifer, near Whitewater creek, raised 20 bushels of peaches. I

was informed that peach trees come into bearing in four years
from the seed. I heard of only one instance of complaint of
winter-killing of peach trees, and that was at Monroe's, 12
miles south of Traverse City and 20 miles from lake Michigan.
Nectarines are raised by Judge Fowler, at Mapleton, on the
Peninsula, and probably at other places. Plums produce pro-
fusely, and are exempt from all insect ravages. I measured a
shoot of this year's growth five feet long on a plum tree in Mr.
Fisher's yard at Glen Arbor. Mr. L. R. Smith, at Elk Rapids,
raised one stem of Early Orleans variety which bore 22 plums,
averaging four inches in circumference. He also raises the
Washington plum. Rial Johnson raised five bushels of plums.
The different varieties of cherries thrive equally well. I saw
flourishing trees on Mr. Smith's place, at Northport; also at
Glen Arbor. Mr. Wm. J. Bland, at Elk Rapids, has a Bigar-
reau cherry tree that has borne regularly for four years. I
saw thrifty trees on the place of E. Pulcifer.

Pears thrive wherever they have been tested. Mr. Smith's
soil at Northport is probably peculiarly adapted to pears, and
they flourish very finely. They do about equally well at New
Mission. I saw good trees also at Glen Arbor, and in the
Whitewater region. Mr. Smith also succeeds with quinces.

Grapes thrive admirably throughout the region—though
wherever I saw them they were retarded in development by
lack of pruning, by excessive crops, and by too much shade.
I saw grapes bearing well at Leland. At New Mission, I saw
Isabella and Catawba grapes ripened on neglected vines in a
situation badly exposed to the sun. Mr. Smith's vines were
literally borne down with their burden of ripe and unripe fruit
in the latter part of October. L. A. Thayer, on the east side
of Torch lake, raised superb Concord grapes. His vines have
been bearing four years. Isabellas ripen early in September.
Judge Fowler, at Mapleton, has matured Isabella grapes for
four or five years past.

This region is the native home of the red currant, the red
raspberry, and the blackberry. Currants are unsurpassed.
Raspberries bear with the utmost luxuriance, either in the cultivated or uncultivated state. I saw ripe raspberries in Octo-

ber, on the Peninsula, growing on canes of the present year's production. The same canes bore green fruit and flowers. This phenomenon is of frequent occurrence. Mr. Tilley, of Leland, informed me that he had, on the last of October, ripe black-cap raspberries growing in his garden, on this year's canes. Strawberries flourish as well as in any part of the world. Mr. Hannah, of Traverse City, informed me that he raised this year 25 bushels from a piece of ground 50 by 75 feet. Mr. Stewart, on the Peninsula, assured me that he could pick strawberries in his fields every day from the first week in June till the approach of snow.

Few situations suitable for cranberries exist, but Mr. Fisher informed me that a marsh along Crystal creek produces them at the rate of 300 bushels to the acre, and he proposes to avail himself of this source of revenue.

The secret of the wonderful adaptation of this region to the production of fruit, is found in the characteristics of the soil and climate heretofore described. It is likely the sandy plains to the south of the East and West Arms of the bay will be found well adapted to the raising of peaches. The region best protected from danger of winter-killing and late spring frosts, lies between the bay and the lake, in Leelanaw county; and yet actual results demonstrate that the peach flourishes, hitherto without drawback, several miles east and south of the bay.

The recent discovery of the admirable adaptation of this region to the purpose of fruit growing, has caused very general attention to be directed to the subject. Almost every farmer is enlarging his plantations. When at Traverse City, on the 8th of November, I witnessed the arrival of 32 cases of fruit trees from the nursery of T. D. Ramsdell, of Adrian. Mr. Mace Tisdale, who had made contracts for this large supply, informed me that he was introducing $4,100 worth of fruit trees this fall.

XIV. THE INDUSTRIES OF THE PEOPLE.

The leading occupation of the inhabitants of this region must necessarily be the cultivation of the soil. Evidently, however, in a country so densely wooded, the duty which first urges itself upon the attention of the new settler, is to effect a clearing. As pioneers generally desire to realize as speedily as possible the avails of their labor, the chopping and sale of "cord wood" has unavoidably engaged a large share of attention; and the shipment of cord wood to Chicago, and its supply to propellers running on the lakes, have become an important branch of business. In November last, the cost of chopping a cord of propeller wood was $1.25, and a cord of shipping wood $1.50. The difference is caused by the greater care requisite in the preparation of a cord which will pass the market regulations in Chicago. Propeller wood was selling on the dock at $3.00 to $4.00 a cord. Shipping wood on the beach was selling for $3.00 a cord; on the dock for $4.00. Freights to Chicago were exorbitantly high; but the state of things was evidently exceptional and temporary. Even at the existing charges for freight, the price in Chicago left a fair margin for profit to the shipper.

Thousands of cords of beech and maple wood, in the haste to effect a clearing, are simply chopped and burned on the ground. It is obvious that two or three potash establishments would save an enormous waste of ashes, and furnish a desirable convenience for the pioneer. I am not aware that the manufacture of potashes is carried on at the present time in any part of the region. It was suggested to me that a man prepared to buy ashes or "black salts," and to furnish in exchange such commodities as farmers generally need, would succeed in doing a profitable business. He should keep potash kettles for sale to farmers residing at distances too great to justify the transportation of the ashes. These kettles would be used in the manufacture of "black salts" on the ground, thus materially reducing the bulk of the article to be transported to the ashery. It is estimated that every acre furnishes from 350 to 500 bushels of ashes.

Another use to which the forest may be immediately converted is the manufacture of maple sugar. This branch of industry is mostly left to the unskillful and untidy management of the Indians. It is estimated that one man can manufacture from 400 to 600 pounds of maple sugar in a season. This, at a season of the year when no other occupation than wood-chopping is practicable is a source of revenue which the pioneer ought not to neglect. •

The manufacture of lumber is carried on only at two or three points, and though over 20 millions of feet are annually produced, it can scarcely be regarded as an occupation in which the people generally are concerned.

The manufacture of bricks and pottery, though not yet established, is destined to become an important branch of business both for home supply and for exportation.

The manufacture of wooden ware of all descriptions might be successfully carried on where the finest qualities of maple, beech, white and black ash and white pine are so readily accessible on the immediate shore of navigable waters.

Sagacious business men have, also, long since suggested the propriety of the erection of furnaces at Northport, Frankfort and other points, for the purpose of smelting the ores of iron from the upper peninsula. The ores of Marquette can now be delivered by railroad at Escanaba, which is only 85 miles by water from Frankfort, and about the same distance from Northport; while the almost inexhaustible forests of hard timber in the Grand Traverse region render it the most desirable portion of the State for the economical operation of blast furnaces.

XV. SETTLEMENTS.

Charlevoix, commonly known as Pine River, though scarcely within the limits of the Grand Traverse region, is destined to become an important point. It is a new settlement, having a substantial dock, a store and several private dwellings. It is claimed that eleven propellers have entered into arrangements for " wooding" at the dock next season. The dock and fifteen acres of land along the beach are owned by the New York Central Propeller Company. The river has four feet of fall at this place.

Antrim City has just been founded by Mr. L. H. Pearl, who has erected a substantial dock and engaged extensively in the sale of cord and propeller wood. The country back of Antrim is becoming rapidly settled, and must soon demand the conveniences of a store and hotel.

Eastport is just founded by an enterprising gentleman of Detroit.

Brownstown is at present a mere fishing station.

Elk Rapids is by far the most important point on the east side of the bay. It was founded by Messrs Dexter and Noble, who have made substantial and valuable improvements—erecting a first class dock, saw-mill and boarding-house, and opening a store, at which the surrounding country is supplied with all classes of goods at reasonable rates. Two hundred barrels of flour were made here in 1864. The Elk Rapids Eagle is published weeckly by E. S. Sprague, Esq. An appropriation of $3,000 has been made toward building a court-house and jail.

Petobego is the name applied to the settlement around Petobego lake.

Whitewater Post-office is located at the mouth of Whitewater creek.

Hoxie and Havilet have a dock at the southeast angle of the East bay.

New Sweden is a settlement, now nearly abandoned, whch clustered around the saw-mill at the head of the East bay.

Mapleton Post-office is located on Sec. 27 T. 29 N. 10 W. on the Peninsula.

Old Mission (Grand Traverse Post-office) is situated near the point of the Peninsula, on the east side, and was the first spot occupied by the white man.

Bower's Harbor, or Haight's, is a wood dock and point of shipment, near the head of the harbor.

Traverse City is the largest and most flourishing settlement in the region. It is situated at the head of the West Arm. It was founded by Messrs. Hannah, Lay & Co., who have erected very extensive and substantial docks for lumber and shipment, and have opened a wholesale and retail store, at which anything may be purchased, from a paper of pins to a steam engine. The saw-mill of this company is one of the largest and best equipped in the State. The place contains also a steam flouring mill which produced 500 barrels of flour in 1864, two good hotels, one or two other stores, a school-house, blacksmith shops, shoe shops, a photographic establishment, and other places of business. The United States Land Office is kept at this place. The weekly Traverse City Herald is edited and published by Morgan Bates, Esq. The population of the place is perhaps one thousand.

Norris' is on Sec. 28 T. 28 N. 11 W. Here is a dock and a saw-mill.

At Lee's Point is a landing dock detached from the shore.

Sutton's Bay is a small village and post-office, with a detached dock for shipping purposes.

Pishawbey-town is an Indian settlement and Catholic mission.

New Mission (Omenia Post-office) is a mission sustained by the Presbyterian Board. A seminary stands here, taught by the intelligent daughters of Rev. Peter Dougherty, in charge of the mission.

Northport is a port of entry, and one of the oldest settlements on the bay. It was founded by Deacon Dame, who removed there from Old Mission. It has a population of six or eight hundred. It is furnished with one good hotel, and several stores. The harbor is a favorite place of refuge for vessels navigating the lakes; and the propellers have been very enerally in the habit of wooding there—the arrivals amounting, as I am informed, to 400 a year. Northport has two good docks—Campbell's and Rose's. At the bight of the bay, two miles distant, is another dock, belonging to Burbeck and White.

Leland is a new settlement, at the mouth of Carp river. It is supplied with a saw-mill, a hotel, a boarding-nouse and store. The place was founded by Fayette and Thies, who own one of the two good dock with which the port is provided. One thousand barrels of flour were manufactured here in 1864.

Thomas Kelterhouse has constructed a dock, and commenced a settlement four miles north of Glen Arbor.

Glen Arbor was settled by John E. Fisher, six years ago. A good dock exists at this port, and Crystal creek affords water-power for a saw-mill and flouring-mill. Another dock has been built on the south side of the harbor.

On the north side of Empire bluff George Aylesworth is constructing a new dock.

Frankfort on the lake is a new settlement. The improvements in the harbor have been already described. The town site is beautifully located on a gentle southward slope, rising from an elevation of 10 feet above the lake to an altitude of about 200 feet in the back part of the town.

Benzonia is a new and enterprising settlement, founded by a Christian colony from Ohio. From a circular issued in 1864, I learn that the place has been selected with great care, as the seat of a Christian community and an institution of learning. One fourth of the entire amount of land purchased is consecrated to the endowment of the college. The church organization is Congregational in form. The sale of ardent spirits and tobacco, except as medicines, is prohibited in the vicinity of the college. The land is selling at prices ranging from three to ten dollars an acre—one-fourth of the profits going to the college. The secretary of the colony is Rev. C. E. Bailey, Benzonia, Benzie county. The president of the college is Rev. J. B. Walker, D.D.

The Carter settlement, in the south part of Leelanaw county, is a neighborhood on the road from Traverse City to Glen Arbor.

The Monroe settlement is similarly located on the road from Traverse City to Newaygo.

Provement is a settlement founded by Mr. A. de Belloy on the narrows of Carp lake.

The population of the Grand Traverse region, according to the State census of 1864, was as follows: Antrim county, 382 ; Grand Traverse county, 2,017 ; Kalkasca, 9; Benzie, 500 ; Leelanaw, 2,389. Total, 5,297. Within the past year this total has probably increased to 7,000, or over. The population of the township of Peninsula, by the same census, was only 479. It is now thought to be 1,000. Reliable judges estimate the accessions to Antrim county during the past year at not less than one thousand souls.

XVI. HIGHWAYS.

The propeller Alleghany, belonging to Hannah, Lay & Co., makes a weekly trip between Chicago and Traverse City, during the season of navigation, stopping at Northport. The lake propellers stopping at Northport afford communication between that place and other lake ports three or four times a week, on an average. Other means of communication, by propellers, are had from Pine river, Leland and Glen Arbor. Messrs. Hannah, Lay & Co. have placed a small propeller—the Sunnyside—on the bay, which, during the season of navigation, makes the round trip daily to all the more important settlements on the bay, and forms a ready, agreeable and most invaluable means of communication from point to point. Her trips have extended, twice a week, as far as Pine river. Besides these means of communication, small sail boats are always at hand to convey the traveler to his destination, in default of other means of conveyance.

Carp lake, besides small boats, is provided with two tugs, which make frequent trips to different points along the lake.

The common roads are of course new, and, except in the oldest sections, more or less imperfect. The beach forms a useful thoroughfare in summer, and the ice in winter. A system of State roads, however, has been put in process of construction, which is destined to prove an important instrumentality in developing the country. These are:

1. " The Allegan, Muskegon and Traverse Bay State Road," running from Allegan by Holland and Ferrysburg to Muskegon, thence by Pentwater, Manistee and Benzonia to Traverse City. (Act approved 12th Feb., 1859).

2. " The Newaygo and Northport State Road," running from Newaygo north by the Manistee crossing and the Monroe settlement to Traverse City ; thence along the west shore of the bay to Northport. (Act approved 12th Feb., 1859).

3. " The Emmet and Grand Traverse State Road," running from Traverse City by Elk Rapids, Antrim, Pine River and Little Traverse to Mackinac. (Act approved 15th March, 1861).

These roads are all in process of construction. The last has been completed to Elk Rapids. The second is open to the Manistee river. The first is in use from Traverse City to Grand Haven. The latter road furnishes the only outlet to the region during the winter months. A stage, conveying passengers, freight and the United States mail runs regularly between Traverse City and Muskegon. A weekly stage runs between Traverse City and Elk Rapids.

Railroad communication with the southern portion of the State is much needed. Land grants were made, about ten years since, to three different companies, who undertook to open communication between the northern and southern portions of the State ; but the difficulties of prosecuting such enterprises through an unsettled region, in connexion with the more recent distubance of the business relations of the country by the prosecution of a great civil war have prevented any of these roads from penetrating very far toward their northern termini. The roads referred to are as follws :

1. The Grand Rapids and Indiana Railroad, running from Fort Wayne, Indiana, through the western part of the State by Grand Rapids and Little Traverse bay, and terminating at the straits of Mackinac. This road runs about 21 miles east of Traverse City, and 14 from Elk Rapids.

2. The Amboy, Lansing and Traverse Bay Railroad, beginning at Amboy, near the southern line of the State, and running by way of Lansing and Saginaw to Little Traverse bay.

3. The Flint and Pere Marquette Railway, beginning at Flint and running by way of East Saginaw to lake Michigan, at the mouth of Pere Marquette river—a point almost directly opposite Cheboygan, in Wisconsin.

The progress made in the construction of these roads is as follows :

The Grand Rapids and Indiana Road has been graded along some portions of its line, and it is promised that 20 miles will be soon completed from Grand Rapids northward. The Amboy, Lansing and Traverse Bay Railroad has been built a distance of 23 miles, from Lansing to Owosso.

The Flint and Pere Marquette road is completed from Flint to East Saginaw, 34 miles. A road under a distinct incorporation has been constructed from Flint to Holly, connecting the Flint and Pere Marquette with the Detroit and Milwaukie Railway at Holly, thus forming a very important line of communication from Detroit to Saginaw. It is said to be the intention of the Flint and Pere Marquette Company to extend their road beyond Saginaw through Midland City, 27 miles, during the coming year. This will carry communication well toward the Grand Traverse region, which is but 125 miles distant.

It has been very properly suggested that the western terminus of the Flint and Pere Marquette Railway ought to be changed to some point within the Grand Traverse region—as Frankfort or Traverse City. This, besides furnishing an outlet to the richest portion of the lower peninsula, would be in a direct line toward Escanaba, at the entrance of Little Bay de Noquet, from which railroad communication already exists to Marquette on Lake Superior. The distance from Escanaba is about 85 miles to Frankfort, and the same distance to Northport. By this connexion the Grand Traverse region would be accomodated, and the southern portion of the State would be furnished with a pleasant and expeditious summer route to Lake Superior.

The practicability and eminent utility of the communication indicated ought to commend it to the attention of the business interests of the State and country.

XVII. CONCLUSION.

The developement of Leelanaw county has been very materially retarded by an extensive Indian Reservation lying in the midst of an active white population. This reservation was made a few months after the first settlement of Northport. It extended from the village of Northport south to township 28, and embraced the entire county as far west as range 13 west, leaving only the small triangle north of Northport as the sustaining back country for that village. Accordingly, though founded under the most promising auspices, a repressive—perhaps we should say an oppressive—public act has deferred for ten years the prosperity of this important point. The term of reservation expires this year, and it is now understood that the land will be speedily brought into market. Mr. Smith, the Indian Agent, informed me that there were this year only 700 Indians to receive their annual payment of $4 each. This payment, as I had opportunity to observe, is at once transferred to the posession of the merchants and traffickers of Northport in exchange for clothing and provisions—a slight offset to the injuries sustained from the reservation. On the reservation at Little Traverse are 1,300 Indians.

A more general and even more serious obstacle to the development of the region is the withdrawal from market of the odd sections reserved for the construction of the Grand Rapids and Indiana Railroad. I found the complaints on this subject universal and emphatic. The reservations for this road cover more than one half of Grand Traverse county, and the entire region on the east side of the bay. It is but justice to the population already engaged in the development of the country, that the injuries sustained from this source should be discontinued. It may be that the only method of constructing railroads through a new country is by means of land grants ; but it is obvious that in this case, the grants have not secured the end proposed, while they have proved of incalculable injury to the region in which they are located. Any continuance of these grants, and any new grants proposed to be made, should be placed under more rigorous stipulations than heretofore, with a view to securing to the regions incommoded by them a more prompt release from the injuries inflicted. The congressional grant to the Grand Rapids and Indiana and other land grant railroads in the State, expires by limitation on the 3rd. of June, 1866.*

The Homestead Act in its practical workings has also retired from occupation many thousands of acres of valuable land. Large numbers of persons, having entered their "homesteads,"

---

* For the legislation respecting the Grand Rapids and Indiana Railroad see Acts of Congress approved 3d June, 1856, and 7th June, 1864 ; and State laws approved severally 14th Feb., 1857, 3d Feb., 1858, 14th Feb., 1859, 15th Feb., 1859, 12th Feb., 1861, 11th March, 1861, 15th March, 1861, 15th Jan., 1862, 2d Feb., 1865, 10th March, 1865.

have failed to comply with the law requiring actual residence ; and they consequently remain unimproved and retired from the market, or the prescribed means must be resorted to for bringing them again into market. These means, with a view to the ample protection of the first claimant, have been made circuitous, slow and tedious. In consequence, men undertake, only in urgent cases, to secure titles to abandoned homesteads ; and such lands are liable to remain a long time without improvement.

At the present time, most of the land lying near navigable water has been taken up. Receding from the shore, private claims become less and less frequent, and disappear, on the east side of Grand Traverse bay, at the distance of seven or eight miles back. In Leelanaw county we find them distributed from shore to shore, with many unoccupied lands interspersed.

The lands belonging to the general government are the even sections within the limits of the railroad grants, except so far as taken up by settlers. After the 3d of June next, the odd sections revert to the government, except in case of new legislation perpetuating the grants.

The State swamp lands within the limits of this region are scarcely worthy of mention. Those formerly held as such must be nearly exhausted in the construction of the State roads.

The reservations for the Grand Rapids and Indiana Railroad are the odd sections where not previously occupied or reserved, to the distance of " six sections in width on each side" of the road, and, where previously occupied or reserved, the odd sections beyond these limits, to any distance within 15 miles. The maps of the company represent their land grant as extending about 15 miles, throughout the entire region. This extension uses up the unsold odd sections nearly as far west as Traverse City and throughout Kalkasca and Antrim counties.

There is no Indian reservation within the region under consideration, except the one already referred to ; and its limits have been indicated. Indians are seldom seen in any other portion of the region.

Notwithstanding the serious drawbacks to the development of this region, growing out of its remote situation, the erroneous ideas of .ts climate and soil, and the injustice which it has suffered from public legislation, it has, during the past year or two, undergone a more rapid improvement than any other portion of the State. There have been entered at the Register's office in Traverse City, since January 1st, 1863, 1,422 homesteads of 160 acres each, making a total of 227,520 acres. In the same time there have been 467 cash purchases, estimated at 37,360 acres. The lands located with Military Land Warrants and Agricultural College Scrip are at least double the cash purchases, or about 74,720 acres, making a grand total of 339,600 acres. This land district extends from the south line of Manistee county to the straits of Mackinac, and from R. 3 W. to lake Michigan. Most of the settlement, however, is around Grand Traverse bay, from R.8 W. to R. 15 W., and from T 21 N. to T. 32 N. These statements were given me by the Register late in October. The entries at the office during November were 12,450 acres ; of which 1,091 acres were purchased for cash, and 1,240 were located with warrants.

Beyond all controversy, the Grand Traverse region offers stronger attractions to capital and settlement than any other portion of the State or of the entire northwest. Even the mighty forest which has to be felled before the farmer can avail himself of the soil, is probably less of a detriment than an advantage. Besides insuring him an inexhaustible supply of fuel, for the labor of cutting ; besides furnishing him with a merchantable commodity in the form of cord wood, upon which he can realize for each day's work ; besides protecting him and his stock and crops from the severity of the wintry blast, the forest itself is a source of food to horses and cattle, both in summer and winter. It is no uncommon occurrence, as I saw in a multitude of cases, for a settler to make his appearance late in the autumn, with little means but his muscle, an axe, a yoke of oxen and a cow. He selects a spot for his dwelling, and while he fells the trees to supply the logs for his cabin, his cow and oxen support themselves by browsing, and the milk furnished by the cow goes far toward the support of his family. Having erected his cabin, he spends his winter in chopping; and, in the meantime, his stock fatten themselves by browsing on the fallen timber, so that they actually enter the spring in better flesh than they did the autumn. I had accounts of this kind from various sources. Mr. Fisher, of Glen Arbor, told me of a pony that escaped from his owner, and subsisted in the forest seven years before he was caught.

A more thorough system of farming is needed, which will be secured when more capital can be applied to the business. A more varied industry is needed; and this also will be introduced as wealth increases and the advantages of the region become known.

Religious and educational accommodations have kept pace with the development of the region. Traverse City, Elk Rapids, Northport and Benzonia have preaching every Sabbath—many of the settlements further back, once in two or four weeks. At Traverse City and Northport the Congregationalists and Methodists both have organizations. There are also church organizations at Monroe settlement, at Glen Arbor, at Whitewater, at Old Mission, at New Mission and some other points. Schools are maintained within reach of every neighborhood. There are at least six school houses in the township of Traverse. At Benzonia is Grand Traverse college and preparatory school.

Access to the Grand Traverse region is had by propellers from any of the lake ports. The numerous propellers all stop somewhere within the limits of the region ; and, by inquiry, it can be ascertained at what point any particular propeller is in the habit of stopping. Those wishing to reach the bay had better not take passage for Glen Arbor or Carp river (Leland); and those wishing to reach the latter places had better not take passage to the bay. Passengers are landed at Northport two or three times a week; and from there they can proceed on the Sunnyside to any other point on the bay. The Alleghany runs once a week directly from Chicago to Northport and Traverse City.

Visitors are cautioned against allowing steamboat captains to persuade them to be landed on the Manitou islands—a frequent wooding place—since great difficulty is often experienced in getting from there to the main land.

www.ingramcontent.com/pod-product-compliance
Lightning Source LLC
Chambersburg PA
CBHW020327090426
42735CB00009B/1432